The MAGUS

Book 1

A Treatise on Natural Philosophy and Occult Science

by Francis Barrett

(Patricia Spencer)

An Introduction (borrowed from an Internet Source)

The Magus is one of the primary sources for the study of ceremonial magic and for a long time was one of the rarest and most sought after of the 19th century Grimoires.

Barrett's Magnum Opus embodies deep knowledge of Alchemy, Astrology and the Kabbalah and has been cited by the Golden Dawn and other occult and esoteric movements as source material.

Written in 1801 in the middle of the Age of Reason, sandwiched between Newton and Darwin, this was possibly the last epoch that a work like this could be composed.

To The Reader
by Francis Barrett

Although we do not, in any point of science, arrogate perfection in ourselves, yet something we have attained by dear experience, by diligent labor and by study, worthy of being communicated for the instruction of either the licentious libertine or the grave student – the observer of Nature; and this, our Work, we concentrated into a focus: it is, as it were, a spiritual essence drawn from a large quantity of matter; for we can say, with propriety, that this little Treatise is truly spiritual and essential to the happiness of man.

Therefore, to those who wish to be happy, with every good intention we commend this Work to be their constant companion and study, in which, if they persevere, they shall not fail of their desires in the attainment of the true Philosophers' Stone.

A Brief Explanation

I have not included an introduction of my own to this text, first, in order to complete the publication in a more timely manner, coming in quickly after the publication of the Doctrina Antiqua by Thomas Burnet, Div. Antiquities, of which I consider this text to be an acceptable companion for the introduction of cosmogony in the problem of metaphysical doctrine and its controlling history over the progress of civilization; as this was the primary topic of the former and other book title by Dr. Burnet.

I have put some end notes to the text, obviously, at the end; and they speak mainly of whatever I have just said. I prefer these comments to be made after the fact of the text, rather than as the primary notion of its presentation. This is because this matter, being somewhat esoteric to me, has been handled in this text as an apropos mark of what has been written already; and thematically, as the opposition of the former idea of metaphysics and cosmogony remaining at odds in the history of civilization since ancient times.

I say, opposition of the former idea, because the former idea is that one that is about metaphysics and cosmogony being at odds in the history of civilization since ancient times and so I wish to form some text presentations which will oppose the idea that this is the only case to believe about the battle between philosophers and cosmogonists, or in the more particular case, magicians comparing to scientists.

I have no use of myself otherwise, to note such sciences such as are these, magical and cosmological and esoteric in the science of strange occult, except that they are interesting to the overall matrix of my general studies in co-linear histories and philosophies between ancient chronicles and lost (and found) scriptures.

Well, I say that they are interesting fact of writ pieces and documents and fragments to be found concerning cosmogony, as the name overall of the discipline of magic involving occult philosophy, because they are the antithetical problem as stated and also a larger particle of the fact of which there is no faith to distinguish history from history as it may be found in esoteric writ or otherwise in the process of recorded chronicles.

The co-linear research domain, if there is one and there should be, is manufacture to the discredit of itself, concerning more than anything else, the production of magic throughout civilized history and its meaning to faith and to culture. There being no established design of credit to its existence, civilization and its record, as well as the credibility of scriptural histories (and legendary histories) continue to fail. So magic is a work, a topic to contend with in the study of history where is concerns the study of faith and the opposition of faith in cultures and societies which become the standard of history themselves; such as the Jews or the Muslims in the regions of the Philistines and the Phoenicians; and for instance; but the examples never cease on a macro and also micro-cosmic scale.

These are all generalities to establish a motive for an interesting controlling theme of interpreting histories and their influence on philosophy and especially the doctrines of faith between person.

So to continue my confession. Short of saying I am studying the textual libraries of lost and found antiquities, which I am, in order to come at a reason for co-linear history and its effects on our practice of metaphysical philosophy today; and since, I have just said that; I prefer also, not to continue on to convince other persons of my hope to find answers in this question, only to have to confess, this is in order to study the relics of histories more accurately and if ever given the chance; and for instance. I mean, only to say I will study it exhaustively on some other level, given the chance.

I don't consider this many persons' topic of quest, journey or crusade; but, as we often find ourselves and the answers to some of our questions, in what other persons have found out about their own mundane interests, I prefer to submit my findings on small antique matters in this fashion. And so I have no more at this time to say, until the gloss interference of the text and again until the very end.

To my fans, Patricia; to the rest, Patricia Spencer.

BOOK 1 in Parts

PART 1

THE FIRST PRINCIPLES OF NATURAL MAGIC

PART 2

THE JEWEL OF ALCHEMY

PART 3

THE CELESTIAL INTELLIGENCER

PART 1

THE FIRST PRINCIPLES OF NATURAL MAGIC

Chapter 1

Natural Magic Defined
Of Man
His Creation
Divine Image
And of the Spiritual and Magical Virtue of the Soul

Natural Magic is, as we have said, a Comprehensive Knowledge of All Nature, by which we search out her Secret and Occult Operations throughout her vast and spacious elaborate domain.

Whereby we come to a Knowledge of the component parts, qualities, virtues and secrets of metals, stones, plants and animals; but seeing, in the regular order of the Creation, Man was the work of the 6th Day, everything being prepared for his Vicegerency here on Earth – and that it pleased the Omnipotent God, after He had formed the Great World, or macrocosm and pronounced it good, so He created Man the express Image of Himself; and in Man, likewise, an exact model of the Great World.

We shall describe the wonderful properties of Man, in which we may trace in miniature the exact resemblance or copy of the Universe; by which means we shall come to the more easy understanding of whatever we may have to declare concerning the Knowledge of the Inferior Nature, such as animals, plants, metals and stones.

For, by our first declaring the Occult qualities and properties that are hid in the Little World, it will serve as a Key to the opening of All the Treasures and Secrets of the Macrocosm, or Great World. Therefore, we shall hasten to speak of the Creation of Man and his Divine Image; likewise of his Fall, in consequence of his disobedience; by which all the train of evils, plagues, diseases and miseries, were entailed upon his posterity, through the curse of Our Creator, but deprecated by the Mediation of Our Blessed Lord, Christ.

THE CREATION, DISOBEDIENCE AND FALL OF MAN

According to the Word of God, which we take in all things for our guide, in the 1st Chapter of Genesis and the 26th verse, it is said.

"God said, let us make man in our image, after our likeness. And let them have dominion over the fish of the sea and over the fowl of the air and over the cattle and over all the earth and over every creeping thing that creeps upon the earth."

Here is the Origin and Beginning of our frail Human Nature; everything being subjected to His rule, or command; one Creature only being excepted, which was to remain untouched and consecrated, as it were, to the Divine Mandate.

"Of every Tree of the Garden you may freely eat. But of the Tree of Knowledge of Good and Evil, you shall not eat of it. For in the day that you eat of it, you shall surely die." (Gen. 2:16)

Therefore, Adam was formed by the Finger of God, which is the Holy Spirit, whose figure or outward form was beautiful and proportionate as an angel, in whose voice (before he sinned) every sound was the sweetness of harmony and music. Had he remained in the State of Innocence in which he was formed, the weakness of Mortal Man, in his depraved state, would not have been able to bear the virtue and celestial shrillness of his voice.

But when the Deceiver found that Man, from the inspiration of God, had begun to sing so shrilly and to repeat the Celestial Harmony of the Heavenly Country, he counterfeited the engines of craft. Seeing his wrath (the Deceiver's wrath) against him (Adam, Man) was in vain, he was much tormented thereby and began to think how he might entangle him (Adam) into disobedience of the Command of His Creator, whereby he (Deceiver) might, as it were, laugh Him (the Lord) to scorn, in derision of His New Creature, Man.

Van Helmont, in his Oriatrike, chapter 42, speaking of the entrance of Death into Human Nature, etcetera, finely touches the subject of the Creation and Man's disobedience.

Indeed, his ideas so perfectly coincide with my own, that I have thought fit here to transcribe his Philosophy, which so clearly explains the Text of Scripture, with so much of the Light of Truth on his side, that it carries along with it the surest and most positive conviction.

"Man being essentially created after the Image of God, after that, he rashly presumed to generate the Image of God out of Himself; not, indeed by a certain monster, but by something which was shadowy like himself. With the ravishment of Eve, he, indeed, generated not the Image of God, like unto that which God would have inimitable, as being Divine. But in the vital air of the seed he generated dispositions; careful at some time to receive a sensitive, discursive and motive Soul from the Father of Light, yet MORTAL and to PERISH.

"Yet, nevertheless, he ordinarily inspires, and of his own goodness, the substantial Spirit of a Mind showing forth his own Image, so that Man, in this respect, endeavored to generate his own Image, not after the manner of Brut Beasts, but by the copulation of seeds, which at length should obtain, by request, a soul imbued Light from the Creator and the which they call a Sensitive Soul.

(By imbibing false consumptives and also re-generating them through their outer organs in order to create the self-generative of the Tree of Knowledge and so to be independent of the Creator, possibly and subsequently, they would initiate mating with beasts to control their Creation and in order to re-create them for their own means. The Originals clearly did not understand the instructive path of the Garden as it was established in Ethic, for Life Eternal; insomuch as they believed the Creator Void and their experiences lacking the determinism of Ethics. Whereas, the entire metaphor of perfection and paradise, discusses the place as a peon, a icon, an institution of Utopian Ethics, which were also so inherent in the Creation, they could not in sapience be denied. Therefore, it was in lacking this creative order for inherent ethics, that their understand would fail. But for myself, as faith, it has been sometime my belief, that, all of the Origin of Sin was an acceptance of a poison creation that could change the order of Creation itself.

But to proceed with the words of Van Helmont in Oriatrike.)

"For, from thence (Eden expelled) has proceeded another Generation, conceived after a Beastlike manner, mortal and incapable of Eternal Life, after the manner of Beasts; and bringing forth with pains and subject to diseases and Death; and so much the more sorrowful and full of misery, by how much that very propagation in our first parents dared to invert the intent of God. Therefore, the unutterable goodness forewarned them that they should not taste of that Tree; and otherwise he foretold, that the same day they should die the Death and should feel all the root of calamities which accompanies Death."

Deservedly, therefore, has the Lord deprived both our parents of the benefit of Immortality; namely, Death succeeded from a conjugal and brutal copulation; neither remained the Spirit of the Lord with Man, after that he began to be Flesh.

Further. Because that defilement of Eve shall thenceforth be continued in the propagation of posterity, even unto the End of the World, from hence, the Sin of the despised fatherly admonition and natural deviation from the right way, is now among other sins for an impurity, from an inverted, carnal, and well-nigh brutish generation; and is truly called ORIGINAL SIN.

That is, Man being sowed in the pleasure of the concupiscence of the Flesh, shall therefore always reap a necessary Death in the Flesh of Sin; but the Knowledge of Good and Evil, which God placed in the dissuaded Apple, did contain in it a Seminary Virtue of the concupiscence of the Flesh, that is an OCCULT FORBIDDEN conjunction, diametrically opposite to the STATE OF INNOCENCE, which state was not a state of stupidity.

Because He was He unto whom, before the Corruption of Nature, the Essences of all Living Creatures whatsoever were made known, according to which they were to be named from their property and at their first sight to be essentially distinguished; Man, therefore, through eating of the Apple, attained a Knowledge that he had lost his radical Innocence.

For, neither before the eating of the Apple was he so dull or stupefied that he knew not, or did not perceive himself naked; but, with the effect of shame and brutal concupiscence, he then first declared he was naked. For that the Knowledge of Good and Evil signifies nothing but the concupiscence of the Flesh, the Apostle testifies, calling it the Law and Desire of Sin.

For it pleased the Lord of Heaven and Earth to insert in the Apple, an incentive to concupiscence; by which he was able safely to abstain, by not eating of the Apple, therefore dissuaded therefrom; for otherwise he had never at any time been tempted, or stirred up by his genital members.

Therefore, the Apple being eaten, Man, from an Occult and Natural Property engrafted into the Fruit, conceived a Lust and Sin, became luxurious to him and from thence was made an animal seed, which, hastening into the previous or foregoing dispositions of a Sensitive Soul and undergoing the law of other causes, reflected itself into the vital Spirit of Adam.

(So therefore this supposes that we may say, that from the Fall of Eden, Adam became the son of Satan, since he became the animal seed of the same and lost his sensitivity to the original facts, or ethics, of creation which had kept him free of his sin. It seems an unoriginal doctrine and one again which I will say soon here, that the sin being present with the co-existence of such a creation as what was Satanic outside of the Garden of Paradise, the entire test was exactly to have the new creation fall into the same sin of the outer region of itself and its influential domain. We have no order to understand that Eden was a place of universal power or influence or physical time or occurrence; it was a creation of a higher power sent to fulfill itself in the greater domain of a hostile and pre-fallen universe. We understand this from the concurrence of the Creation Myth with the Creation itself.

The idea of the perfection and the sin itself remains elusive in this explanation of the facts of the lives of human creatures when they were found together in a pristine state; but this incurs the greater problem of conceiving without fault of accidents of any kind or sins, that the pristine order has a nature of paradise conceivable to seem natural to human conception in the dualist order of Satanism and Holiness under the same Heaven.

I've explained my beliefs about the original later; but I would also concur that if we could not in the first place speak of the Satanic co-existence with Paradise, we could not explain sin in a theological sense, so we must explain that the fallen state is a larger mystery than what we might conceive to ourselves; or else if it were not more involved in explanation to us, we should be void of the temptation for Satanic poison to tempt away the qualities of purity and holiness by the original state of their being inherent in us. And we could not therefore (without the pre-concept of co-existent in dualism) discuss the intellectual and physical faults incurred against a better original in the actual state of falling from the primordial state to the co-existence with Satanism. In order to become so imperfect as the story would have us, we must have a greater reason than the temptation of nothing from nothing; or in the temptation of purity from impurity; this would not seem an adequate temptation. But I say the Creation, in Eden, was already fallen into the dualism of the original sin doctrine and therefore, the test was to prove it in co-existence with a fallen universal from which the original fault was already made.

Or, in another sense, I myself, do not understand that the condition of original impunity for damnation from holiness and purity in Creation, is satisfied in this depiction of judgment stating that it is a temptation by Satan to fall to sin and concupiscence with the fornications (or whatever diminutive and unsatisfying ideals would be) of materialism and its opportunity for failing sins redundantly and never-ending to the extent that the point cannot be made without the original in example; whereas, the original in example is not actually given, given the fact that, Satan is Universally in co-existence with Creation, outside of the Garden Gate, so to speak.

It is child's play to have to understand what kind of evil scheme would produce such a judgment; no less innocence or ignorance could achieve such a simple delusion as to believe that the entire impunity is in the realization of the dualist and poison fable as belonging to a litany of the same in fallen histories and that therefore, this evil scheme is by far gone from judgment value, but to establish a termination or at least a repose from its damnable and conventional standard of all creative and original principles abstracted to that end.

Such an evil scheme as to raise up an exemplary shekinah to an opposing evil overlord and proclaim dominion over his existence but to continue to set up the domain of his fearful worship, is such an evil scheme as intended to show itself for the co-existence and fallacy it should be understood to be.

So arguably, if there were some opportunity for perfection in the dualist state given, it was only for Immortality in the Dualism of Evil together with the Lord's domain in the Earth; as we see He recedes it progressively to the promise that, the Doctrine of Sin established was a code dishonored for a false tradition at failed and recreant self-creationism; there being already a Master of the Trade, or Masters, what more would the faithful want to achieve but to be false men? Therefore, the failure is for Immortality in a Coeval Existence; and nevertheless, a lesson is granted in the Universal Scheme of the Original Dualism which is unrevealed, unless it is as simplistic as the ideals put forth of a poisoned creation.

Leaving aside more speculations, what was failed, again, was the exercise of an Immortal Faith in the dualist rationale of Creation, against the Immortality that was granted in that place; and that, there was no apparent reason for that Man to obtain it. But as for this being the actual statement of the spiritual man's perfection at origin, it fails other theological precepts.

The Creation Myth of Eden, definitely has a certain theological argument to state that it is a necessary doctrine in order to establish the expulsion of sin from perfection and the need to conceive of the spiritual power inherent in mankind's Soul to achieve a heavenly concept of spiritual perfection.

(Without which ideals, there is little to spare in the order of our mortal beliefs in themselves for redemption from sin; they are truly all a waste of time.)

So we must test ourselves and understand the test of original sin or sins, to underestimate that Creation in Eden stands for the dualist nature of the Universe and Eternity and that Utopian facts of the Tree of Knowledge were and are differentiable in abilities to be perfect and spiritual from the original concept of heavenly purity.

So to continue on my original course and to close (being myself also damned never to achieving the doctrinal point about the precedent and co-existence facts of the argument which I try to elucidate further to the end here.

The fact that Satan exists anywhere is under the domain of the Lord's powers in the Prime Order of the Universe; it is within his original compulsion towards Paradise as the Prime Order for a new Immortal Creation and for His order of perfection towards them.

And nevertheless, we understand from philosophy that a Utopian model of life is not perfect or paradise and therefore, I say we understand these persons to have been already fallen and so forgotten and we ourselves the same. So the test was met to recurring the co-existence of dualist powers in the Universal Creation and this is then the acceptance of all things of doctrine. The mystery remains in whatever was precedent; how could it be known, being so unsatisfactory to reason and those reasons outstanding for the purpose of employing natural poisons in order to recreate in an obviously perfect domain. That should seem heretical. Does it?)

And, like an ignis-fatuus (an in-cognate; a deception or pathological delusion; or an otherwise secret combination (of logic); an unknown and mistactical aesthetic), presently receiving an Archeus or ruling spirit, an animal idea, it presently conceived a power of propagating an animal and mortal seed, ending into life.

(It is presently my belief concerning this Allegory for the Statement of Original Sin, of Fallen Man and Fallen Creation; or concerning otherwise the Mythology of Life Created from Matter – the Creationist Mythology of Eden – however anyone will name it; that, the logically inconclusive hardware of the mythology presents itself precisely in order never to be answered without some higher faith in a higher creed of scientific litanies of order; an order which includes this higher Master of the myth itself. This would seem to be the Patriarchal Construction of the Earth and its place of inhabitance in the Universe, as being resident within the idealization of a Savage and Brut Society of Creatures, not unlike animals in the ending statements of their primordial mythologies whether they have been established as true or not to the commanding presence of Eden.

    The lesson to ascertain in this Hardware and Construction of the New Eden Myth (of scriptures and chronicles), as I've called it, is that we must believe first and foremost in our own Creation Mythology and not in any course of prior mythologies which would lead us into the fallacies of co-existing dualisms with evil; and we must undertake the consideration of this lesson in order to be saved from the brutish savagery of the instinctual human mind to become thus as bestial and animalistic. The redemption from dualistic creation, that which is created both good and evil, is the redemption of the conceptual demand for spiritual sin; sin against any faith in a prior order of true perfection. So as this human mind and creature is redeemed by the observance of his own situation in the primordial order of creation, despite the dualistic and pre-original (then) fellowship (with Satan), then it is revealed that the creature status of the prime order was prefect, and that after the first instance of dualism, the creation of mankind is raised up out of a state of fallen virtues from the original prime order. (Satan has no order to originate himself independently of any of God's Creation; but that he does and survives the brutality of the ages against Judgment of his own fault, he recreates himself again as the master of the domain of Eden. This fact makes it impossible for mankind to survive a test he would not in the first place.

But Satan, was created with the origin of the species of man as he was in a separate congregation of the ages; and remaining in the fallen precincts of the world and in the universe at large, he became the potentate of Jehovan Mythology, the prototype of dualistic creationism, to the expulsion of the concept of existence without a prime order of evil (or, the Tree of Knowledge, being good and evil). Were this an original fall for creation, such a dualism could not ethically exist. It was in the second case, in the case of his lacking creation, that he was the catalyst for the fallen state of mankind's final demise from immortal precepts with the word of holiness in his origin; now the creature status must suffer bestial recurrences to obtain his freedom from the brutal and senseless factionalism of fallen ages unbeknownst to himself.

So I continue, more than probably pointlessly, with the original intention for the species of objections to the final doctrine of the Eden Myth as being determinant towards any real progress for the revelation of the origin of the universe and as possibly being the principal of determining science in creationism; and I say this, not because of its mythology, but because of its lacking principles to complete its own realism as wanting to conclude a satisfactory situation of rules for the order of trials and judgments as I have stated my argument ahead (or before); and so therefore, also, the Myth itself does stand as educational to our existence for the performance of a compulsion to endeavor in matters of conscience most certainly and more definitely the same with a dire need for determinism to do it.

In which case again, the original doctrine of perfection, purity and immortality was in a prime order shared with a bestial universe and that, lost in sense, with only immortality left to them to gain by the determinism to be a separate mythology, even the re-assertion of faith to ideal perfection by the Creator himself, could not redeem the first lesson of perfection, from the beastial fellowship of evil in the prime union of creation; that whereas it had not been in the first place, it was again in the second, joining Satan in the Ages.

But let me go on.

The matter is entirely embroiled and unreasonable.

It cannot be extracted by human reason and this is to avoid the reasonable codicil of the responses of prior Creation Mythologies. Students of Divinity report that they are sometimes taught of a school of thought which states that there was never a Higher Creator who ordained that there should be a mistake of creative representation in the order of Eden or any place; that there was never an Eden but only a perfect creation. And therefore, if mankind were to return into the place of his primordial ancestry he would find himself among unexplained mythologies, and unexplained ideologies, alien and abstract and perhaps even so foreign that the atrocities of the Talmudic judgment may seem the more perfect rationale of a conditional and perfect response.

But concerning the ideals of truth, were these things all true? – they were also all poison idols by themselves, standing out, as imperfect representations of other myths; and referring to whatever are "these things being true," viz. to, a bestial primordial ancestry without any place of Eden for even intermediary refuge; were it also holding so true, that, there was never a separation of singular purity into dualistic evil. The Entire Construction of the Eden Myth again falls short of any merit as the original answer to our questions of origins, because of the primal state in spiritual perfection and the lacking theological importance for any rules of dualism in the aesthetic concerns of a perfect order that are becoming mythologies. The perfect order must be held theologically true and not as mythology, which is the reverse order of fantastic faith today from the empirical; though gnostic inquiries continue on into the archaeologies of Paradise and the co-linear Ages. And there also has no meaning offered in the suggestion that an original doctrine of the kind is besides meant to be the very point of instruction for warfare with dualistic faith; and no more; faith being more important still.

So the commandment received by the Tree in existence with Paradise, was the Knowledge of the Prime Order of Existence and its fallen state, a-priori of creation and that it had become a ritualistic form, and traditional practice of evil, for the universal worship of creation itself by poisoned fables and lacking ethics of purity or holiness or original perfection.

The commandment was consumed and consumable in the form of a pre-fabulous poisoned fruit; the seed of all passion, at the expulsion of reason from its Zenith of perfection in Zion.

The Knowledge therefore of good and evil thereof, was an imperfect knowledge of an imperfect system of fallen virtues, whatever they had been and of those things standing episteme in refusal of pristine concepts. So that, a better creation was that which was brought about by poisoned rituals and rites and those which continued on condemned from ages unknown. But anyway, this is only a certain school of thought. There are more expressly immaterial philosophies on the matter of poison apples and bestial monsters and their influence on primordial values of perfect order and creation.

It is only after these considerations that the question becomes, if the Lord of Heaven himself established the ritual lesson of poison annihilation to his Creation as an admonishment to beware of His presence, identity and covenanting and to become unaware of false stratagems for self-creative philosophies as the other Creationists who will have destroyed themselves by this prior theory. And not even as any posit of an ultimatum besides eternal damnation outside the gates of hell awaiting judgment for the mistake? Should it not at least have been to have said, the mistake was pre-existing and outstanding in the debts to immortal nobility therefore? Or truly this miracle of creation is falsely a cheat of theology to acquaint itself with a fallacious leader of philosophy named God and not the Lord, Creator of spiritual means of powers. But he is only Satan and Jehovah, in diligence with evil aiming to control what was already set up for judgment.

Did the Lord of Heaven intend for the Creation to Fall with the remainder of the Fallen Universe? Was it ever the choice of the new Establishment as the oldest antiquity in name of the Garden of Eden; the Oldest New Establishment. Was the Garden at the mercy of choices outstanding besides Immortality; is Immortality the only zenith of perfection? May we ever believe in ourselves to be creatures of prime order and reason and spiritual as well, unformulated from material errors?

Because if we could, we could believe in a co-linear and related antecedent to the Talmudic Authority which controls our systems to the end of the critical data of our own global environment and the extinguishment of our possibilities to decipher any historic truth besides, and returning to the Lesson of the Origin, of the offending and un-cipherable myths? But we have no authority of our own dominion in reason to ever be greater than the controlling myth of ages, to say, even complete a spectral order of magic without self-extinction as a race of believers of some witchcraft or other; that would be to say, to underestimate mythologies as magical sciences. This was the original Satanic fault. And having nothing to believe in the magical possibility of spiritual reality and its aesthetic reasons for existence, as realism for creationism, then we have fewer hopes to cipher the magical and actual codes of histories riddled with the combinations of incomplete myths; we are Satan's children; intent upon purifying a Jehovan redeemed order in order to establish our own Utopia at the defiance of the past ages. But how could we do it without the final closure upon the trial of evil with magical principles having no reasons but fables?

It is a nearly pointless endeavor. As some theology students are sent to learn for themselves, to endlessly consider the wit of a nearly perfect human creation to have even originated of its own grave judgments of self-extinguishment for a past they never knew or exercised the rules of, but to exist in the slavery of fabulous dogmatic trades of persons who themselves were myths – could have some other meaning besides the fall of philosophical hopes into the barbaric and this alone being the admonishment. Has it seemed to have taken effect for us? As theological students there is merit to say that the rules of the edifice of these lessons, being a church or a temple, it is establishment of rationale to continue to try to obtain immortal concepts; this establishes the prime fact that, these objectives were never true in a former sense; but that barbarisms themselves created their choice in reason.

This finally must recall, given the attribution of mortal thought with the hope of immortality in a spiritual sense, that an offer of formal logic precludes the matter of an a-priori motive; or otherwise, there is no motive or a-priori matter; and the offer of logic to obedience was an offer of a logical failure to the standard of repose in perfection. This is the recounted point, which is obviously also the theological point of the doctrine of sin and redemption from it; that the barbarism of past mythologies accounted for no redemption from a fallen state of barbarism, but continued in its demised statement. The primal order of perfection must be the matter of logical offer to satisfy the demand for a hope of immortality; and it is, but only in the spiritual state, after the fact of fallen barbarism, compounded with a Satanic (or that is, mortal or fallen) Creation in Eden. This brings up the final question that, surely the Tree of Knowledge was in itself a fallacy of the mythology and that the alternative Tree of Life was that which was the law of the perfect order, determining for some people that, the Tree of Knowledge establishes an Eden set to fail for the premise of the New Covenant of redemption (for the barbaric ages).

I have said this all enough different ways. It has seemed in some relevance to this text however, because the idea of perfection and the responsibility for it in the primal order of Eden is that which drives persons to insane notions of barbarisms in the first place, concerning the magical doctrines of science and their establishment in the distinction of theological treatise. The difference of all codes of science, theology, magic and primal perfection and order, have no realism all together in the option for barbaric solutions with imagination given the origin of the myth and its ages of failure in a-prior besides as mentioned (that is was Satanic interpretably). So I am slightly adverse to presenting this point of view in theological introduction to the magical doctrines set forth in this treatise, to which I am able to understand some importance and relevance to all ages. This is because it has all faith toward the end of fallen Eden and its original pristine condition and I say otherwise to wit of the end that the magical discipline in faith is that same one which destroys the theological concept of spiritual perfection. And it is one which should be considered a promise to the ages in revelations.

So I have objected to some of the determinism of "our parents" being those to have distinguished that we have no magical concepts to use in the reason of our science. I believe this is the pre-existing faults of barbaric ages and the determinism to rationalize barbarisms with mythologies to the end of unexplainable results in phenomena.

But it must not be taken too lightly however, that besides in theology, but also as a point in philosophy, the ideal instance of being complete and then choosing an incomplete alteration to one's own dimensions in favor of following a model of corruption which gleans as a zenith but only in a metaphor, is rather, the same as believing a fallen faith in original has a pristine order of beneficial concept of fellowship with one's own goodness; and one also which assumes that self-annihilation and creative absorption is a primary goal of reasonable creatures. The entire assumption is fallacious to all systems of reason. But as for our original sins, those to forsake the ideals of immortality for a pre-original and unshared system of mortality and its fables; they are forgiven, theologically and by the orders of philosophy to underestimate these rules of writ; only in so much as we have a path, or undertake a healthy pathology which attains a point of reason against the cost of all other false attainments which are unreasonable, which are the orders of fabulous sins in the first part and before their consumptive promises are fornicated.

So anyway, we conclude the diversion of treatise and doctrinal objections with this objectification for all cause and effect, once and again, that the prime order of the world, Eden, was all a mythological determinism on the parts of the principal movers of the story, to eradicate the Creator in the first place and replace him in the second. The influence here is clear to any person who has read any of the mystical documents of the gospel pseudipigrapha and that it was the aliens who were appointed at first to protect the original creation as agents of Satan, for and against him to hurt the objectivity of those players of the Universal stage; but those influencers are they who are entirely lost in the Eden Myth and who appear in subjugate writs.)

Furthermore, the Sacred Text has in many places compelled me unto a perfect position, it making Eve a helper like unto Adam; not, indeed, that she should supply the name and room of a wife, even as she is called, straightway after Sin, for she was a Virgin in the intent of the Creator and afterwards filled with Misery. But not, as long as the state of Purity presided over Innocence, did the will of Man overcome her. For the translation of Man into Paradise did foreshadow another condition of living, than that of a Beast. And therefore, the eating of the Apple does by the most chaste name, cover the concupiscence of the Flesh, while it contains the Knowledge of Good and Evil in this name and calls the ignorance thereof, the State of Innocence.

For surely, the attainment of that aforesaid Knowledge did nourish a most hurtful Death and an irrevocable deprivation of Eternal Life. For if Man had not tasted the Apple, he had lived void of concupiscence and offspring would have appeared out of Eve, the Virgin, from the Holy Spirit. But the Apple, being eaten, "presently their eyes were opened," and Adam began lustfully to covet intercourse with the naked Virgin and defiled her, the which God had appointed for a naked help unto him.

But Man prevented the intention of God by a strange generation in the Flesh in Sin; whereupon there followed the corruption of the former Nature, or the Flesh of Sin, accompanied by concupiscence neither does the text insinuate any other mark of the Knowledge of Good and Evil, than that they "knew themselves to be naked," or speaking properly, of their virginity being corrupted, polluted with bestial lust and defiled.

Indeed, their whole Knowledge of Good and Evil is included in their shame within their privy parts alone; and therefore in the 8th (chapter) of Leviticus and many places else in the Holy Scriptures, the privy parts themselves are called by no other etymology than that of shame; for from the copulation (alternately, intercourse) of the Flesh, their eyes were opened, because they then knew that the Good, being lost, had brought on them a degenerate nature, shamefulness, and intestine and inevitable obligation of Death; sent also into their posterity.

Alas! too late, indeed they understood, by the unwonted novelty and shamefulness of their concupiscence, why God had so lovingly forbade the eating of the Apple.

Indeed the truth being agreeable unto itself, does attest the filthiness of (the) impure Adamical Generation.

(Later, Adamic Generation; an eternal and succeeding generation bent upon self-creative and material principles without any faith for immaterial opportunities that might occur naturally without magical occurrences – the damnation of the self-creative (the concupiscence as he names it), concerns not the original statement of perfect creativity and whatever it was, but the exclusion of those principles for others chosen that were ordained to be true by a separate authority, a material authority for a material creation and not a spiritual and perfect creation; so the assumption of the Higher Spirit's ordination of that more ancient rule of objectivity of Eden, that Eden itself was the perfection of a credible and attained Zion, was destroyed in favor of hedonism and replacing the need for spiritualism. This is the objectivity, as often described, of the Adamaic Generation. That Zion had no perfection or reality that was not in material, hedonistic recreation of the original principles of spiritual completeness.

Also secondarily as a side note, many scripts of the various pseudipigrapha suggest and chronicle treatises and exegetical creation myths, as well suggest, that the Deluge was the eternal cleansing of the Adamic Generation and that it was replaced, notably without any mother in particular of Legend, by the Brahamic (Brahman) Generation, which also was messily brought into the judgment by the Captivity of Babylon and the imminent closure of any universalism implied against the judgment, by the completion of the expulsion of man from the exegetical gardens of the old testament, by the remnant civilization of the Sumerians in the intra-testament period before Christ and the remainders of the imperial societies in co-linear writ upon the same Earth, appearing literally to have been "expulsed" from the opportunity of survival by natural cataclysms and apocalypses.

I have early on in my studies made the messy reduction of the (Ancient) Brahmans to Abraham in descent, for the purpose of associating a co-linear theory even there; so that in the latter day Tribes of Israel after them, they are those the same who had traveled to Egypt and were saved by Moses. It is unnecessary to anyone to agree with my estimation of the Brahmans being so named as antiquated descendants of an ideal sort of Methuselan Abraham, since it also lacks any merit of reliable histories; but I use it all the same to present ideals and assert it the same, as the principle of which, the Ancient Patriarchs must prove themselves influential to the end of Ancient Sciences in the temporal universe and in celestial matters. So many people have felt that the same eastern faith owes himself to the original Patriarchs and especially Abraham.)

For the impurity which had received a contagion from any natural issues whatsoever of menstruations or seed(s), and that by its touching alone is reckoned equal to that which should by degrees creep on a person from a co-touching of dead carcasses and to be expiated by the same ceremonious rite that the text might agreeably denote, that Death began by the concupiscence of the Flesh lying hid in the Fruit Forbidden; therefore, also, the one only Healing Medicine, of so great an impurity contracted by touching, consisted in washing: under the similitude or likeness thereof, Faith and Hope, which in baptism are poured on us, are strengthened.

(For the record, this explanation of impurity, not so much eludes my wit, as it does not exactly; but as it does not present itself as anything but a liturgical answer finally to the conclusion of the matter of an impure sin according to the pattern and pathology of an established conduct and logical consequence, cognitive of it. But the penalty of the sin is the conclusion of a logical consequence; whereas, the sin itself must be larger in importance, than that of a logical consequence; and this objection, seeming by the time of ages, inconclusive to any proof of a meaning, is not held relevant by most to the idea that it is merely a logical consequence; that is, a logical consequence of using an illness or ill way to prove a medicine for health itself and being a consequence of punishment, logically in consequence, equal to the punishment of a sin; whereas, the punishment of a sin, per se, should exceed the limits of ordinary consequences and concern the problem of a commandment which has been broken, not merely a material law.

So to go on with my convoluted argument, objecting to the differences between sacred sins and material "sins" and that possibly broken material laws do not qualify for these same judgments for sins; we continue to have a larger basis for obtaining magic thereby, because of the larger spectrum against the diversities of actions, against the diversities of restrictions; by refusing to include that all actions in Eden was a sin, whereas but only the false objectification was a sin.

But to continue my argument with myself, objecting to all usage of the original sin ideals to the requirements for obtaining magic; obviously, I am wrong to say so, because much of an entire theological doctrine of sin attributes the worship of idols to original sins, un-sacreding worship, and therefore, we need to begin wherever sin begins, which, in doctrine, is in Eden. Whereas, the dialectical arguments of creationist myths overall do argue what I am saying, that, magical sins, original sins and sins concerning those decisions to be free of covenants against false idols (or false reason), is pre-disposed to the occurrence of Satanic Evil in the Universe, prior to Eden. So the original sins against idolatries, are pre-disposed, and also not as directly influential to the practice of magic, as would be the actual implication – that, idol worship is entirely without worship in the practice of magic; (this means that the doctrine of original sin by this reason, does expressly imply that there is no worship to combine magic with idol worship; that IS the entire condemnation; so it is that I assume we are beginning with Eden in this universal prospectus; but, continuing as I was meant to in the first place.).

This is a topic of punishment as sin, not so much the idea that, Satan was the influence of evil in the Garden of Eden, which I have established is an opinion of doctrinal consecration to ensure circumspection to the later matter of universal idol worship.

So I return to the mundane issues still concerning me with all this – what has to be necessary pre-disposition of respect to religious doctrine for the practice of very controversial universals in the applications of magic. Here I begin again.

In the more serious issues of defining unreasonable causes for actions, it is more interesting to determine if the pre-cognitive matter had never any material source; in which case, the objective creativity of a cognitive cause was at fault and therefore, in that sense, any objectivity of the new cognitive substance was an impurity from the start; that the parents of the creative matter believed it a purpose of intercourse was subjective of a choice to themselves; alternately mythologies are written in which the choice of a creative material cognate unformed from the prior point of logical demesne is otherwise a matter of a weapon or tool which has an impure purpose; more likely.

But for the sake of the argument of the liturgy that we must behoove ourselves to learn; considering that it is in the Garden of Eden and by its purpose to become all undefinable mythology, for later science to wonder at; that magic is expulsed from faith and science and philosophy, and nevertheless, magic is the argument of all of those disciplines themselves for the matter of proof, then, whatever some accomplished shaman of the rite could define, we must listen if willing and pursue to some aspect of understanding. The judgment here is all for sexual immorality in so much that this conduct was for the purpose of destroying the Creator.

I contend, were this the argument alone and not for the attainment of some other civilization of which they were meddled to learn, then surely poison and madness prevented their understanding from attaining the ritual dogma necessary to overcome such a false instruction as to use some extra means to encumber life itself and its miraculous course already. And this all, provoked upon the Garden luxury and gift, in order to destroy the Creator.

This being the obvious goal to my wit, it falls short to me that such a thing could be rationally installed to credible action and outcome and storied belief, with the actors as such of those who are faithful, and also Satan and the Lord of Heaven; this is a doctrinal establishment as noted. So being obvious that the storied report establishes that the goal by ignorantly adapted Satanic rituals, is achieved for the purpose of eradicating the Creator and establishing the Garden as a Colloquium Region of a greater Universal, obviously interpreting of themselves to be the ambassadors of themselves to the other Ages; and not merely for the enjoying of lacking circumspection with concupiscence, as he says, in so far as, the creature status has no realism for understanding the emotional back-biting of such a course of action. Why would they have hoped for greater enjoyment what they already enjoyed, if it were not that enjoyment was the obvious goal to remain pristine as the superior gods of their own realm; they obviously understood themselves to be imperfect and lacking enjoyment in the zenith of the Universe. The tragedy is realizing at the end, that they were the chosen people of Zion, the Zenith of the Universe by verse. (And this is what, as a completely unrelated aside, enjoys me to understand that this Zenith, this insolvable location of Eden and Zion together, was Mars; by the epic poetry maintained to Paradise and Mars for Ages of wit to this folklore. But that is beside the point; also because of the planets' newly reaffirmed forbidden domain; as the wasteland opportunity of a cataclysmic finality. Let us forget my opportunity for advertisement of mundane concerns and continue on with the research of our magic. So lastly.)

The Genitors of Satanic Guilt considered themselves falsely made without enjoyment, the objectification of another race of watchers. This is clear in mystical treatises. I have no more now to add to everything I have spoken of the former concerns, however, there is more in later concerns to be seen.)

For as soon as Adam knew that, by fratricide, the Firstborn of Mortals, whom he had begotten in the concupiscence of the Flesh, had killed his brother, guiltless and righteous as he was; and foreseeing the Wicked Errors of Mortals that would come from thence; he likewise perceived his own miseries in himself; certainly knowing that all these calamities had happened to him from the Sin of Concupiscence drawn from the Apple, which were unavoidably issuing on his posterity, he thought within himself that the most discreet thing he could do, was hereafter wholly to abstain from his wife, whom he had violated. And therefore, he mourned, in Chastity and Sorrow, a full 100 years, hoping that by the merit of that abstinence and by an opposition to the Concupiscence of the Flesh, he should not only appease the Wrath of the incensed Deity, but that he should again return into the former splendor and majesty of his Primitive Innocence and Purity.

But the repentance of one age, being finished, it is most probable the Mystery of Christ's Incarnation was revealed to him; neither that Man ever could hope to return to the brightness of his Ancient Purity by his own strength and much less that himself could reprieve his posterity from Death; and that, therefore, Marriage was well pleasing and was, after the Fall, indulged unto him by God because he had determined thus to satisfy his Justice at the Fullness of Times, which should, to the Glory of His Own Name, and the Confusion of Satan, elevate Mankind to a more Sublime and Eminent State of Blessedness. From that time Adam began to know his wife, viz., after he was 100 years old; and to fill the Earth by multiplying according to the Blessing once given him and the law enjoined him – "Be fruitful and multiply."

Yet so, nevertheless, although Matrimony, by reason of the great want of propagation and otherwise impossible cursory Succession of the Primitive Divine Generation, was admitted as a Sacrament of the Faithful; if therefore, both our first Parents, after the eating of the Apple, were ashamed, they covered only their privy parts; therefore that Shame did presuppose and accuse of something committed against Justice – against the intent of the Creator – and against their own proper Nature: by consequence, therefore, that Adamical Generation was not of the Primitive Constitution of their Nature, as neither of the Original Intent of the Creator.

(And they were adjudged as Mortals.) Therefore, when God foretells that the Earth was not of the Primitive Constitution of Nature, as neither of the original intent of the Creator; therefore, when God foretells that the Earth shall bring forth thistles and thorns and that man shall gain his bread by the sweat of his brow, they were not execrations, but admonitions, that those sort of things should be obvious in the Earth: and, because that Beasts should bring forth in pain – should plow in sweat – should eat their food with labor and fear, that the Earth should likewise bring forth very many things besides the intention of the husbandman; therefore, also, that they ought to be nourished like unto brute Beast, who had begun to generate after the manner of brute Beasts.

(The inimical actions taken by the pair to supersede the government of Eden that the Lord established over the human person of Adam (as also over his wife, Eve, and to be definite over any other power, such as Satan) as Himself, the Head of Creation, could have been then, to be sure, that act of itself to return them to the dust of the Bestial Creation, here noted and the same one that preceded Eden, so to continue to plough and toil and struggle without heavenly rest, both literally and figuratively, (that is, physically and spiritual) and without any divine concepts but those attained through personal trials. Concerning Eden, the doctrine of ploughmen's shares to mortality, is the distinction between Immortality and Mortality; though through the use of other fictions, either religious, scriptural or mundane or otherwise, the distinction of mortality from immortality, is a concern of spiritual knighthood and ephemeral understanding which exceeds the cause and effect of mortal limitations. But in the folklore of Eden, immortality was stopped because mortality required that mankind, through Adam, would observe the material laws of the Universe in the Earth and forever; until he could redeem himself in the Heavens.

That the cognition or recognition of these acts (being those of a ploughman or laborer of any kind, however esoteric), as belonging to the bestial kingdom in their understatement and execution remains in the later times after this folklore an elusive objectivity; what could not incur the wrath of needing work, even to think of it? So to assume (or presume) whatever immortality stood for in the Age of Eden, the Lost Time of its Place, is to undertake the existentially impossible without ridiculous imagination. Feasibly this admonition should have been enough to dissuade the pair to incur some judgment of the kind upon themselves. And for modern man or later ages to their own, there is no account of a judgment to be made, even because the entire script is a wonderland for imagination: to live in Paradise without a single care for need or for aging or for death? Has such an edict have a meaning for us? What are the actions of enjoyment without the problem of sins or deadly sins to suffer in any case of judgments? And were it so in anyway true, why would there remain the secured outage of a natural enemy? Would it be to exist without cause of conflicts with an enemy? Again and again; the script establishes a pre-script which establishes the easiest end of super-humanity and immortality to mortal men, who may have been demagogues.

Let us consider ourselves amongst perhaps the demagogues of the ancient philosophical world and without the problems imposed upon two persons such as Adam and Eve as specific. And were this the same domain for the fortune of the scripted situation; would they so find themselves, everyone from Aristotle to Seneca, so enticed by the production of evil in the recognition of Satan that they should feel it their inherent creation to destroy Him by destroying themselves? There is no answer to any of it. Many people would enjoy the imaginative play of both Aristotle and Seneca becoming immortal and continuing on in their god-like leadership of the Universe at large without more recurrence to setbacks in the millennia of civilizations. But deadly sin learned in that Golden Age of Eden's mythology and folklore, forbids such a consideration besides for the script of even greater compositions of mythologies; stories which, notoriously these groups of demagogues, would not cast themselves for fear.

With no inspiration in model set ups, we return to the pair.

Their methods and attributes denied the more important sequence of knowledge inherit in the puritanical Creation of Origin; and therefore, the bestial is that quality which determines the brutish specimen of Mankind thereafter, and probably more than indistinguishable to himself. Or this is what Barrett suggests.

To the end of our own mortality as liberals of the Eden Myth, unable, more than usually to obtain its essence for a pragmatism of faith, there is no human person born without complaint of the matter of the bestiality of the creative mind and its lacking graceful principles at origin of education; we consider ourselves a brut species from a former god-like example of folklore, because everything obtained to individual possession, is not only chosen but studied and curated; or, more succinctly and exactly, brutally won; whereas, we are to understand that, all of life's attainment was a granted gift of special enjoyment and fulfillment (in Eden). These are the promises to the redeemed of Heaven (thereafter) and only for the moment of achieving Heaven; whereas also, because the attainment of Heaven is considerably one of the most brutal testimonies of any faithful people in any time of any kind to undertaking, it is usually the experience of choice rejection for the fact of its ridiculous and folkloric requirements of a brutal and impractical work experience. So therefore, this is all accountably also to ourselves, if we so wish to believe, because of "our parents". Whereas, I say it is owing to our willingness to continue to promote Satanic idol worship of any kind to ourselves; but I only say this as a contrary doctrine about the guilt of these parents; their circumstance, therefore, may still be those to blame by origin.

It is only semantic falsehood, if not at least, less than compline and more than uncircumspect, to assume that the Higher Divine had no understanding of the better creation and also the artifact (of the creation) than the ideals of idolaters and philosophers of self-creationism throughout all the scrutiny of the ages. (For that matter; how could there have been idolaters and self-creationist "experts" in the pristine environ, to be the forbidden fruit, unless it were that they existed and co-existed elsewhere. Some people and say they know, that they did in fact co-exist elsewhere; charlatans of ideals   and vendors of the same scriptures, whoever they were; or practitioners of idol faiths at least; the world histories are filled with the ideas of such civilization. So why should the pristine couple endure ridicule of being artifact of the Lord's creation? Because it was a Divine Creation ordered by Divine Authority; whereas, there may be other kinds of co-existing peoples thereof the Universe, of which no one will ever know, except by more legends about unfallen graces. Should there be such a place for the brut race of a knighted purpose to Satanic will? Of course not; the expulsion from Eden determines everything to be the same of its own design and Nirvana to be a hellish fairytale, possibly also the basis for the forsaken goals of Paradise.)

    The matter of grace, particularly, Divine Grace, concerns the bestiality at all points of spiritual warfare in the human mind and it is one which we find out to be false knowledge and false semantic, only by the trials of reason and/or faith; therefore, anyone can understand themselves at that peril. However, it is never the more obvious attribute to assume the rational success of a perfect and divine outcome of circumstance in anything mortal. This loses the zenith point of epistemology on all points, including those of faith, that the aesthetic reason inherent in anything created, resides independently of the virtue of human interpretation. (So that, all human interpretation, even this one if you will, is void of a perfect institution with the origin of truth and so that, the hope for divine truth or anything sublime in the nature of reason and/or faith, is subjugate to the ideals of a more perfect interpretation all the time; how is that kind of reason obtained. Only of course, by rejecting all the false idols of reason which were obtained at the origin of opinion. So deadly sin is all a false opinion. This observation, at last, I consider an important one to make here, because we speak of the practice of magic. And if it is not then self-evident why human interpretation and its errors is at fault of its need to decide upon objectivity that serves itself, then we begin at fault of hoping to determine a science of magic.

The argument is thus; that the origin is lost and this is by the act of introducing actions termed "sinful" into a prospectus existence termed "divine"; and for argument alone, sinful and divine have no co-existence in consequence of their co-operation if it is in extinction of the creator of the existence which is divine; Divine Heaven and sinful consequences are not in co-distinction and therefore, one must be omitted from the other; so if we are unwilling to understand such a distinction there is little hope that we need to try ourselves with it but to become a part of the co-incidental relationships of experiences to the probable mystery or uncertain psychology of sin and its mythology and co-existence with science. Since, I understand from the doctrinaire of the mythology that the written word of human interpretation for the ages itself has given that, there is no more to interpret about the folklore, except that it is the determinism of the word of sin at the origin of its entry to the logical divine. That being the point, furthermore, to reintroduce a sinful logic, an a-priori rationalism to the point of execution, into the practice of magic, is to reintroduce the forbidden fruit of knowledge to its point of execution with a new machinery. I believe this is all the admonition we must take from this lecture; and my own; to avoid the fornication motive of the use of magical precepts.)

It is likewise told Eve, after her transgression, that she should bring forth in pain. Therefore, what has the pain of bringing forth common with the eating of the Apple, unless the Apple had operated about the concupiscence of the Flesh and by consequence stirred up copulation.

(As another aside point; and for whatever derision I incur that will be said as I understand; I prefer (having no other way to state it) "intercourse" in use as the word hereof for the word of "copulation" which Barrett uses and re-uses with respect to his "concupiscence" (rather as to say, indolence and insufferability) argument, which I reject; and the distinction I make of the preferred word replacement, concerns his "concupiscence" argument per se. Though it may not be an appropriate syllogism, as he hopes to use it, it might be a distinction to make that a "copulation" in his object is not the same execution of betrayal as the other term I suggest. So he argues it. An act of copulation sometimes refers to more bestial circumstances than the other act as said; as in and most typically, the idolaters' wheels which copulate clay. The usage of the term copulation is sometimes used instead of the other, to make also the distinctions of beastial inclusions in the mating ritual; whereas intercourse holds fewer opportunities of the same semantic inclusions as ritual. So this would help to understate whatever these copulations were about; because in more vulgar myths, the exact same accusation was that which made the gods of the origin of the material world, failure to the completion of superior strength against greater forces than themselves; they were in effect the first idolaters. Is it to say that the Edenic residents were similar to the larger demagogues of earlier folklores? Then they were created with the same Satanism in mind. So I prefer the other term.)

And the Creator had intended to dissuade it, by admonishing them from the eating of the Apple. For why are the genital members of women punished with pains at childbirth, if the eye in seeing the Apple, the hands in cropping it and the mouth in eating of it, have (not) offended? For was it not sufficient to have chastised the Life with Death and the Health with very many Diseases? Moreover, why is the womb afflicted, as in beasts with the manner of bringing forth, if the conception granted to Beasts were not forbidden to Man?

After their Fall, therefore, their Eyes were opened and they were ashamed; it denotes and signifies that, from the filthiness of concupiscence, they knew that the copulation of the Flesh was forbidden in the most Pure Innocent Chastity of Nature; and that they were overspread with Shame, when, their Eyes, being opened, their (both) Understanding saw that they had committed Filthiness most Detestable.

But on the Serpent and Evil Spirit alone was the top and summit of the whole curse, even as the privilege of the Woman and the Mysterious Prerogative of the Blessing upon the Earth, viz. that the Woman's Seed should bruise the Head of the Serpent. So that, it is not possible that to bring forth in pain should be a curse; for truly with the same voice of the Lord is pronounced the Blessing of the Woman and Victory over the Infernal Spirit.

Therefore, Adam was created in the Possession of Immortality. God intended not that Man should be an Animal or Sensitive Creature, nor be to be born, conceived, or to live as an Animal; for of truth, he was created into a Living Soul and that (much), after the True Image of God; therefore he as far differed from the Nature of the Animal, as an Immortal Being from a Mortal, and as a Godlike Creature from a Brut (or Savage).

(They were reduced to savagery proclaims the repeated incident of the use of the term Brut, according to Barrett, since, Brutes in ordinary histories are either legendary of archeological, known to be indigenous Savages. Perhaps, for the reason of their archeology, truly unpromised, but for the mundane arguments, they should be sought in the influential archaeologies of prehistoric savages. There might be found the key clues to a lost Edenic tribe amongst the origins or the Mesopotamian result of other histories; or otherwise the Ganges as precedent to that location of Ur, what became of the exterior garden. Though Savages pre-date them by at least 8-10 thousand years at their end; the Savages, the conclusion of time to Eden has no meaning except to be put in precedence of all the writ of the scripture, so predating Job and preclusive of the cataclysms spoken of in primordial Creation Myth accounts accounted for sometime within this 8-10 thousand year gap since the Savages disappear. In order for the scripture to have its authority of account, it must not be producible either through time reduction or time assertion; but in general; so, generally, the Neolithic. The Bronze Age is specifically Mesopotamian in historic writ well-known; so probably too late. But this is an interesting distinction for the finding of fact of legend; that, there being no archeological hope of it, whatever it is, there could be some intermediate indication for whatever the hope is; in that case, we see the distinction of the planet itself (discrediting my Martian ideal, but so to go onward) within the disguise of the composition of the planet itself, without notice of the place itself. Could this be true, then their time were lost to the geology of itself and only perhaps. So that, we could incur that the new savages were in the Holocene of the Neolithic Age, since, by encyclopedic instruction we find out that the Holocene is a "quaternary" geological period beginning approximately 12,000 BC and including ourselves today and the Neolithic ends at the height of the over-lapping Bronze Age of the Egyptians. The Hammurabi Code takes the rule of Mesopotamia but not of Egypt in the time of the Bronze Age and so it is also reason to believe, it is too late to find Eden there. But as for my own assertion of Mars; to answer an outstanding question; I would suppose if Mars had to be in the definition of the Earth found; I forget these things for metaphysics, then I would put it in the Pacific Plate of the Neolithic Age with the Halocene Geology. In which case, this is devastated by finding no Halocene Geology on Mars. It doesn't present itself a riddle to be answered by metaphysical amateurs of physical histories.

(I've chosen the Pacific Plate because of its circular plate construction, so as to inhabit a dome, first of all; and second of all, because it is a fluid construction of means from the original land masses since Pangaea, or it deconstructs easily as a separation thereof; and lastly, because the Pacific Plate associates the more true link to the hollow earth and/or flat earth idealisms which could be the clue to the core inconsistency – or, lacking magnetic fulfillment of some sort – which allows for continents to arise from plates over billions of years; in which case, this being the more interested view of these latter days of continental development, it is to say that the entire planet was destroyed from a larger mass of itself.)

Let us now make it the more outstanding endeavor to relate ourselves to the completion of this chapter in its original.

But I do submit that the ideals presented in this text of "occult", but nevertheless, of natural science, does present a metaphysical ideal which needs to be substantiated and that is, wherever might the physical occurrences of this outer region domain entertain itself through the Earth, in which the ordinary means of the ancient sciences which determine magical extenuation into the physical known; it seems to place itself somewhere into the cosmogony of earliest creationism and so an understanding of the relation of this newer cosmological physical from the ordinary physical needs to be accounted for. As we have said, some attributes of responsibility for ourselves seems also the order of good conduct. And all together, this discusses the need to believe in a civilization that permeates the Ages and is also endorsable by this mythology here; so archeology or at least geology and should there be a discipline in the same interest of metaphysical necromancies such as these concerning the gaseous resolutions of planet domes, then also that to add to the objectivity.)

I am sorry that our schoolmen, many of them, wish, by their arguments of noise and pride, to draw man into a total animal nature (nothing more), drawing (by their logic) the Essence of a Man essentially from an Animal Nature. Because, although Man afterwards procured Death to himself and to his posterity and therefore may seem to be made nearer the Nature of Animal Creatures, yet it stood not in his power to be able to pervert the Species of the Divine Image.

Even so as, neither was the Evil Spirit, of a Spirit, made an Animal, although he became nearer unto the Nature of an Animal, by hatred and brutal vices. Therefore, Man remained in his own Species wherein he was created. For as often as Man is called an Animal, or sensitive living Creature and is in earnest thought to be such, so many times the text is falsified which says, "But the Serpent was more crafty than all the Living Creatures of the Earth, which the Lord God had made."

(This is for that and) because he speaks of the natural craft and subtlety of that living and Creeping Animal. Again, if the position be true, Man was not directed into the propagation of the Seed of Flesh, neither did he aspire unto a Sensitive Soul. And therefore, the Sensible Soul of Adamical Generation is not of a Brutal Species, because it was raised up by a seed which wanted the original ordination and limitation of any Species; and so that, as the sensitive Soul in Man arose, besides the intent of the Creator and Nature.

So it is of no Brutal Species, neither can it subsist, unless it be continually tied to the Mind, from whence it is supported in its Life.

Wherefore, while Man is of no Brutal Species, he cannot be an Animal in respect to his Mind and much less in respect to his Soul, which is of no Species.

(This very nearly asserts that the Man's Soul created is tied to a Mind of no determinant Creation and that the Soul in itself is kept with that very spiritual guide of sinful thought (or pre-cognizance) and action, who holds the Satanic realm; furthermore, what is not Satanic in the accounts of brut savagery and their means for civilization; it equates the same meaning of essence to deny a higher spiritual power which determines an exit point to salvation of its own means, as the credible determinism for realization? But in the savagery of the earth, the decision is to procreate with the basest of objectives in order to accomplish all the effects of the demands of brutish determinism; even animalistic, however unrealistic. Histories are filled by these speculations of Mankind's refusal in his physical, archeological origins, to realize that his spirit had no physical determinism outside of reason, whereas, the savage would say the only spiritual reason was in physical determinism. To the former statements; theologically, the Soul has no material creation, however, it is retained in paradise (without further corruption, that is, without further involvement into the material mystery) in a spiritual state and as the accomplice of the greater spiritual identity of this Mind-Life machinery which defines all of creation to the mythologies involving it as God and Spirit, the Central Soul.

But it is a rather freakish idea to this perfectionist schema, to introduce a physical Soul back into the dispensation of a conscience cleared conscience of Sin. So I do not understand that there is a lack of Creation, theologically implied to the Creation Myth, concerning the involvement of Man's Soul. Whereas, this interpretation for us designates that Man was independent of his need to seek out the meanings and intuitions of his Soul in order to be at peace with higher concept; he had no Soul, because he was part of the Eternal Soul. This is rather pantheistic and very demagogic for the most part of its assertion and it continues to want to say that these originals were in fact the epic story we know from Greek Mythology of Martian perfection in the ages of the Sun. So these were sun-worshippers, similar to the Egyptian Ra Priests. The mysterious implications are never-ending of course.

But about the Soul of Man; it remained uncreated by Sin and unattained of hellish knowledge, which would expatiate any divine or immaterial conception from take flight into fantasy. Whereas, this statement here, makes the Soul created and uncreated both as the perpetrator of the Mind's false thought. I am willing to say that, leaving all of Creation to the Lord Higher of Heaven, it would be His own compulsion to decide these thoughts in men's Minds, however, it lacks the principles of a grace in theological purity; which was the fault in the first place. Was Eden a theological society? Then it was doomed by the problem of heretics foregoing; that is the only point.

Furthermore, it seems impossible that the place ever could have had a domain in relation to the edifices of the plates of the Earth known inclusive of its supposed time in millennia; but that it must have belonged to the Ages and this is because of this extemporaneous attribute of the Soul suddenly, which theologically has no domain in literature of credible authority; Man's Soul belongs to cosmogony if it is fallen and otherwise magically belongs in Nirvana or its equivalent until it is forsaken. There is no such Eden to be known of this doctrinaire or gentile convenience that we need not concern ourselves with the sentient ability of the spiritual mind to avoid its body and live in peace with its education for spiritual graces.

It has no domain; Eden has no credible domain on this basis.

For it did, the Principalities of it found out by finding the physical forces of the physical Soul (as we understand Satan to have done to Job) and so the fact of this history remains the plague of all Ages and Aeons. So I declare it for myself to be thus. But to also involve the idea that the Soul of Man was also created in this laboratory expressly for Judgment Day and its accountability, is also a teaching of false theology to the allowable doctrinaire of the present times after Eden. The Soul of Man, throughout the physical ages was in doctrine created in the laboratory, if need be, of Heaven; and, descending into the reasonable ephemeral realms of peace, it found itself void of a body and so sinned, in which case, it was removed from its native home and restored to Heaven to await the Judgment of its actions when found out to be a sinner in its descended place of the Firmament in Nirvana.

But this Nirvana does not also satisfy to be the place of Eden, because it is and was fallen before the fact of Eden; Nirvana preceded Eden; and Nirvana was already a place of dualistic ideals, to the extent that it was (and is) perfectly safe from the imperfection which allowed it to fall in the place which it did, which was a temporal and imperfect realm descended from Heaven into the Firmament; and also because, there, there was never the name of Satan, because he was expulsed also from the Firmament, able by that place to accede again into the higher halls of the Heavens and was thrown to need to re-create himself in the outer region of Nirvana. That there was never a Satan in Nirvana by doctrine also needs to be examined, as it is a false statement intended to confuse the spiritual or those who hope to be; because by his name in Nirvana, there is none, but by his identification, there is the purified form of the same fallen angel among many of the serving angels lost there in the purgatory for Heaven. Nirvana is by indictment thereby feasibly understood for the Firmament for all the quality we understand of it. (As the Sophia also explains all these rules of his keeping, though we continue to confound the identity of Satan expressly with the Devil for the translators' continued determinism for it in favor of the mischief of the works of fallen angels so using this mistake. But it is in the

Nirvane Firmament, the Pristine Universe of expelled and deserted Heaven itself, that Satan receives a new crown and identity to prove for himself there the consumptive power of his own hopes and assertions against the Lord of the Higher Realm. For there he could evade the higher angels and achieve the reasonable terms of his own creationism and assume the powers of all the surrounding Natural Universe and Animistic Universe left to him, for which he fought; and so this all would destroy his name of evil Ascension and lend a theme of rebirth into a Supra Generation, even as the fallen scriptures before the Eden Myth do promise was their case of understanding. So I believe it, that even Satan ceases from the threat against Man's Soul as it is protected and created, except against his own determinism and that he is powerful thereof enough to become even those kinds of God's which were known to the former creationists, or even a kind of Eternal Zoroaster, or Leviathan, but anyway, an animistic and alien sort of angelic person who has become the demagogue ruler of the Firmament. We understand by this all, if it is plausible to our own imagination, that the Devil who is the thief of souls and fiend of faith, is someone else, who has no admission to any term of redemption.)

    Even so, that would say that Eden was Nirvana, which cannot be; unless somehow it can. It becomes therefore, a never-ending phrase for Judgment Day to find one's self at the end of it there, wherever it is, still wondering about the mystery of this myth; so much for the problem of time in cosmogony and if we have allowed magic to influence us therefor; obviously not.

    So then, the last problem that Eden, according to some historians, states that, feasibly, Eden must have occupied this domain of fleshy angels of the lowest estate of Heaven, this Firmament, of my asserted claim belongs to Satan's domain, as if Satan were some kind of redeemable Set and for instance; though he would burn in a fiery furnace this only distinguishes himself to his own faults, from Christ, who fell to the lowest estates of Hades without guides of Chaos or Mystery. But for the claim of the Judgment of all Eternity and whatever we may not understand falls to those who are outside of the realms of creation by that day, Satan is not figured in, but remains returned to another later time, as Leviathan. The scribes   under orders do annihilate him like Judas, into the preparatory fires of Hell in order to say that the Devil is lost there as well; but he is not, because, there is at that time, still no recourse for the completion of Salvation and only Satan himself has paid the Devil's debt, without recompense of the debt itself, that then obviously must be paid by some other misfit. What we owe to Satan is the fault of our creation in Heaven and what we owe to the Devil is the fault of our soul, lost with him in Hell. This is because the very act of dualism against the full power of the Lord of Heaven, is that kind of re-creationism which destroys the foundations of reason from that estate into the things of faith and deniable further then by mutable statements; and so this is Satan to thank for mutability and dualism. However, to live in Eternity for the Judgment of the Soul's second death there, is the work of a creator who has no plan but to consume the creative soul in his own fire without re-creation under the Heaven, besides himself. If that is now all of what we conceive, as a-priori evil and determinism, then, we are making a very bad clerical mistake which is evil in itself, for it again assumes the higher power of the Creator, to have no meaning in his perfect ideals. He has no discrimination in his own Covenanting laws and their severity, since these are well known theological accountabilities for grave punishments in the spiritual domain of the Soul.

But we know of truth of the reason involved in faith, that the Soul of Man was not given to Hell for the Judgment Day for the price of Sin, unless Man himself would give it by volunteerism to the Devil's charm and to this very Day of obligation; and this all, given the fact of Messianic Creed in substantive realism of the words of the compulsion to understand the terms given of covenanting; so that, a spiritual protection is always implied in the covenant, in which case, all rejection of it, describes the volunteer terms; whereas in so far as the temporal realms outside of the Infernal Region is conferred, the terms of covenanting continue to require the rejection of the furnace itself; what does that tell us? That the requirement of the special covenant of protection over Man's Soul, is in the very edifice of the new world of Satan's determinism and not to avoid it with a spiritual intent to no longer demean the personal matters of faith.

Therefore, the expulsion from the Primordial Garden was finalized by the intention to inhabit the Soul as well with a better Creation, a subsequently hellish dwelling for the Devil's charm.)

Therefore know, that neither Evil Spirit, nor whole Nature also, can, by any means or anyway whatever, change the Essence given unto Man from his Creator and by his foreknowledge determined that he should remain continually such as he was created, although he, in the meantime, has clothed himself with strange properties, as Natural unto Him from the Vice of his own will. For as it is an absurdity to reckon Man glorified among Animals (this is what pastoral mythology was finally about which became the aboriginal idea of paradise for the ages, especially in polytheist cultures); because he is not without sense or feeling, so to be sensitive does not show the inseparable Essence of an Animal.

Seeing, therefore, our first parents had both of them now felt the effect throughout their whole bodies of the eating, of the apple, or concupiscence of the Flesh in their members in Paradise, it shamed them; because their members, which before, they could rule at their pleasure, were afterwards moved by a proper incentive to lust.

Therefore, on the same clay, not only mortality entered through concupiscence, but it presently after entered into a conceived Generation; for which they were, the same day, also driven out of Paradise. Hence followed an adulterous, lascivious, beast-like, devilish Generation and plainly incapable of entering into the Kingdom of God, diametrically opposite to God's Ordination by which means Death and the threatened punishment, corruption, became inseparable to Man and his posterity.

(Another statement for the record. This expulsion from Eden (scene), for myself, has been an inspiring ideal in co-linear studies between scriptures and geographical histories of legends, into which I understand the idealic lands of legends must have been born; whether they were true in some realism or not, but nevertheless true representatives of Emirate civilizations (or cultures) to those of the Earth, including most of those we have derived from the assertions of Celtic and Druidic and Savage Tribes throughout the Ages. Their relationship to the Earth, seems in fact to be a separation from Eden either in prior or in postern or in both; does this not lend us some insight into the terms of the intercessory God having been a greater power, an Overlord in the polytheistic system of the Brahmans or Orientals and also the intercessor of the Ages, as God was styled often in the earliest pages of writ, by the Jews. Even Joban domain is rather legendary. And so, and for instance, I style all of the lands of the Expulsion of Eden, wherever it might find itself in the Universe, as the location of this Idyll and Fairy Forest of the Cosmological Writers of the Renaissance and also the basis for Albion and the other lands of Myths established in the course of "other" explanations for Man as an intellectual Creature descended of angels or sapient angels. Should we find ourselves out as self-creative, we should surely chastise ourselves nonetheless for having no grace towards ourselves and furthermore, for having no grace towards our imagination and its need to exist with some aesthetic meaning outside of the toil of objectivity without achievement.

 So I offer it as a cognitive error, if you will; because it may well be; that seeing the leaves of the tree that Odin put in a Garden and considering this a sign of Immortality sent into the Earth even to the last ages of prediction, that besides so saying, such is the land of demesne to be sure of. As I hold no influence outside of the Avataristic scheme and to be a little more predictable for what I mean, should I have been Eve, we might hope to remember these words with respect to what is obvious about the spectrality of light in our Avataristic and mutable bodies. I am joking. I would never allow that we should become creatures of absolute flight into the fiction of a legend, as it was the Creator's original design of graduation and no one else's we tamper with, so why trouble Him more with the vanity of His hopes? We should become the compulsory objectives of the primary order of dominion; Man, as Avatars. (I make the same mistake twice to be more insignificant.) And, this demise   into the inane is all in vain to be cupid with my own meaning. Or else they were all characters of a lost Epic tale similar to the demagogues of Greece including Trajan and Cupid himself, since, I cannot make anymore promises to carve out the look and the meaning of the location anymore than it has to be true to us today, being buried by millennia of other moments of inconclusive vows to and for Satan, excepting some knighthood avoiding him through imaginative epics.)

Therefore, ORIGINAL SIN was effectively bred from the concupiscence of the Flesh, but occasioned only by the apple being eaten and the admonition despised; but the stimulation of concupiscence was placed in the dissuaded TREE and that OCCULT lustful property radically inserted and implanted in it.

But when Satan (besides his hope and the deflowering of the Virgin, nothing hindering of it) saw that Man was not taken out of the way, according to the forewarning (for he knew not that the Son of God had constituted Himself a surety, before the Father, for Man), he, indeed, looked at the vile, corrupted and degenerated Nature of Man and saw that a power was withdrawn from him of uniting himself to the God of Infinite Majesty and began to greatly rejoice.

That joy was of short duration, for, by and by, he likewise knew that marriage was ratified by Heaven – that the Divine Goodness yet inclined to Man – and that Satan's own fallacies and deceits were thus deceived. Hence conjecturing that the Son of God was to restore every defect of contagion and therefore, perhaps, to be incarnated. He then put himself to work how, or in what manner, he should defile the stock that was to be raised up by matrimony with a Mortal Soul, so that he might render every conception of God in vain.

(This explanation deludes most theologians and should at least unless there be some puritanical objections which should prove themselves better, that the presence of Satan from the beginning of the Institution of the Creation as a Garden of Eden, includes the preclusion of a test and nothing else of the decision to avoid Satan for one final fault against him or not to and I understand the judgment to have been that Satan should remain in the judgment and this concludes the mystery of the Original Doctrine of Sin; that the Creation returned to creative Satanic means, after some greater cataclysms noticed in the co-linear accountability of records, and that he could no longer consider himself eligible of more supreme and ultimate forms of human attainment (those things formerly defined in aesthetic as divine or pure in the a-priori of the sentence), without considering himself in a magical sense without reason or faith to save him.

Whereas, he did not require reason or faith to save him in that original statement of his creation; he had something divine and pure to account for whatever must remain the explanations of this higher realm of none account to our material universe. This merely sends creation back to the point of the Ages and Aeons and into the Universe, complaining and unprotected from magical ideals, as we do find ourselves. So the Tree was a Tree of Magic, I surmise for my own enjoyment of the faith. Which was confounding for the purpose of an explanation for itself suddenly appearing in the Holy Precincts. Or had it a rationale for existence, having no point in joining the semantic ideology of a more distant relationship to a determinism of a past.

It was poisoned by some Satanism or Archonite in a prior age of creation. I could not have been in the image of the holy creation of the times and otherwise must be a specter of death to a beast or alien, because I cannot conceive myself, who would choose to live in a temporal ideal, without some Utopian means of dominion over the bestial of course and nevertheless, in objectification of living among its litanous forms of knowledge and explications of it. That this demands for a Satanic society is also a conclusion I make which I find therefore upsetting in itself and so I promit that I make all these conclusions too simply but so does theology today and for centuries and that is because, it must stand to prove the Ages of Creation since a prior co-linear era of intolerable cataclysms.)

Therefore, he, Satan, stirred up not only his fratricides and notoriously wicked persons, that there might be evil abounding at all times; but he procured that Atheism might arise and that, together with Heathenism, it might daily increase (I would prefer the word Hedonism, but I have no surety in the matter of atheism and heathenism in concert for Hedonism); whereby indeed, if he could not hinder the co-knitting of the Immortal Mind with the sensitive Soul, he might, at least, by destroying the Law of Nature, bring man unto a level with himself under Infernal Punishment. But his special care and desire was to expunge totally the Immortal Mind out of the stock of posterity.

("Therefore, he" (Satan); the text here changes Satan immediately to the Devil as became the vogue of the latter half of the last millennia and was not the final word of ancient mystics; I also make objection here – as a student of more ancient Christian myths – and understanding Satan to be no less than Leviathan, to including that Satan and the Devil are the same identity, so I prefer where we begin with Satan we remain with Satan and where we begin with the Devil we end with the Devil; there being practical objections to make; that the Satanic continues in its own combinations to abjure a world without end that is in hedonistic terms again Heaven, but that the Devil himself abjures the position of Hell as the fact of primordial and original existence and that until all of creation is returned to the primordial mass that serves as such of his tyranny, it has no meaning of respects to his objectivity. The Satanic Ideal (at least in literature) is simply in objection to the creation of Heaven in the material universe. So definitely against the appearance of Eden. Many persons believe that it was the accessory of the Devil's power to Satanic powers existing unjudged throughout the Ages that the Devil Incarnate achieved his mysteries of deadly powers with sinners in the Occult. I agree with this constituency. Satan is a fallen angel of a school antithetical of the supremacy of Heaven and the Devil is a monster against all Godly conception and creation. To account the motives of the creature status of the Devil Incarnate no less with the primordial powers of the prehistoric ages in the creature status of Leviathan, which is entirely rejected in realism and materialism today and rendered the magical ideal of metaphysical imagination, without account for influential legend, is to give up the entire Creation to the Beast of Burden and probably also Satan with it. Leaving Heaven a new domain for angels and aliens without conception or existence except in the metaphysical literatures of pseudipigrapha.)

Therefore, he, Satan, stirs up, to this day, detestable copulations in Atheistical Libertines; but he saw from thence, that nothing but brutish or savage monsters proceeded, to be abhorred by the very parents themselves; and that the copulation with women was far more plausible to men; and that by this method, the Generation of Men, should constantly continue; for he endeavored to prevent the Hope of restoring a remnant, that is, to hinder the Incarnation of the Son of God; therefore, he attempted, by an application of active things, to frame the seed of Man according to his own accursed desire.

Which, when he had found vain and impossible for him to do, he tried again whether an imp or witch might not be fructified by sodomy; and when this did not fully answer his intentions every way, and he saw that of an ass and a horse a mule was bred, which was nearer akin to his mother (a satyr) than his father (a demagogue, or a serpent race); likewise that of a Coney and dormouse being the father, a true Coney was bred, being distinct from his mother, only having a tail like the dormouse; he declined these feats and betook himself to others worthy, indeed, only of the subtle craft of the Prince of Darkness.

(Essentially, this is the ancient curse of Set upon Creation, concerning what remains unjudged of primordial myths and their polytheist sciences at large to the later Ideal Ages lost to them. So Set is Satan and also a Serpentine Creation, a Leviathan, a myth of giant beasts becoming giants and destroying the future of any creative intelligence by the terms of superstitions and lacking leadership besides of any philosophy which was not polytheist and deterministic in the use of matter against itself and also against life. This is the state of the fallen ages as often related in legends. So we understand possibly that Satan is one of these fallen gods, not merely a fallen angel and his creationism corresponds to his superintendent myths. Actually and realistically, we might believe this; especially if we believe he may be understood for Leviathan. But the bestiality of the Devil corresponds to the logic of primordial chicanery and mischief with any reason or system of order or even of the domain of Chaos and Hades; he is the outlaw of all conception and all creation and demands an end to Heaven without his creature status; to what end is this? Satan demands the exclusion of Heaven from the domain outside and the Devil demands that the inclusion of Heaven precludes the exclusion of all other creatures not created in his image – as a satyr essentially.)

Therefore, Satan instituted a connection of the seed of Man with the seed and in the womb of a junior witch, or sorceress, that he might exclude the dispositions unto an Immortal Mind from such a new, polished conception; and afterwards came forth an adulterous and lascivious Generation of Faunii, Satyrs, Gnomes, Nymphs, Sylphs, Driades, Hamodriades, Neriads, Mermaids, Syrens, Sphynxes, Monsters, etcetera, using the constellations and disposing the seed of Man for such like monstrous prodigious Generations.

(These satyr-like creatures, were not what I had in mind as monstrous; however, it continues on in keeping with the story of a pristine government by the Heaven being compromised by a corrupt and unjudged history endangered by the Chaos and incongruity of its own fall. So this Eve, becomes very much not unlike the Lavinia of Brutish Histories and her consequence, as a Creation Myth of the primordial lands of the Holy Grail, the Albion and Suri which subsequently became the Red Lands where the grail was transported for eternity; and this is in keeping with the general ideal of the location of Eden in the lower lands and deserts of the Mediterranean east of Greece to its end and continuance again. So the idea is to understand that a prior creature status was in existence but suffering with so much cataclysm and extinction by Chaos as the means of its logical survival that there was already no distinction between mythologies, creatures, their sciences, aliens, and beasts. So we find and that the continuation of the meddling of these creatures like beasts, satyrs, essentially being the mythological purpose of a changeling from a man to a beast in the first place of a man (or woman); these changelings were associated to a Serpentine cultivation for a story of evolution and its mythology in heroic and epic creationist myths. But even theosophy claims this mythology in judgment of Creationist means, as having been made by the same God who ordained the lands and decrees of Eden. What is interesting very finally is that the Devil ideal versus the Leviathan ideal (or Satan in name), is the optionality of the satyr form of a human man (or woman), a person.

And that this changeling is the creation for which the demonic powers are warring for control. Does it not automatically, from the division of Chaos from Hades which we establish today must be separated throughout Eternity from Eden, this theologically, determine that Mankind must suffer a fall from the precept of what is not perfect? Eden is not perfect, but an allegory to some perfection that surely failed even to the end of the a-prior failures of the greater stage?)

And, seeing the Faunii and Nymphs of the woods were preferred before the others in beauty, they afterwards generated their offspring amongst themselves and at length began (their) wedlocks with men, feigning that, by these copulations, they should obtain an Immortal Soul for them and their offspring; but this happened through the persuasions and delusions of Satan to admit these monsters to carnal copulation, which the ignorant were easily persuaded to   and therefore these Nymphs are called Succubi; although Satan afterwards committed worse, frequently trans-changing himself, by assuming the persons of both Incubii and Succubii, in both sexes. But they conceived not a true young by the males, except the Nymphs alone. The which, indeed, seeing the sons of God (that is, men) had now, without distinction, and in many places, taken to be their wives, God was determined to blot out the whole race begotten by these infernal and detestable marriages, through a deluge of waters, that the intent of the Evil Spirit might be rendered frustrate.

(This description now, and we should for the most part, being any kind of mystic or slightly lost theologian (in a library of favorite co-linear sources), -- understand that the demonic world was that which must be judged for its conduct and philosophies from "the start"; a beginning and a preceding Eden. And that the demonist issues were those in confounding the nature of evil for the Later Ages (which we estimate conclude with Noah's accomplishment) and thereafter to the fate of philosophers and theologians alike establishing a restitution and salvation clause of demesne over scientific progress on Earth.

    This should be known by the commingling of stories and legends which preclude any mystery of good or any mystery of evil to the exclusion of the other in the Satyric Creation Myths which we find conveniently to any of our hopes of entertainment to the end of times and they are as old as oral antiquities and speak of the primordial fictions of creationism and universalism themselves; in more adult fictions, they become exactly the challenge of all faith in existence themselves and this is more than any other reason, because, the demonic flavor and character of any legend in the respect of fables and mythologies, has to contain in it, the writ of some promise which establishes for philosophy, whether entertaining or not and whereas, the Devil is confounded with Satan, there is no meddling agreement; but a determinism for the altered primordial creation hell-bent since the mythical time of Chaos' expulsion from Hades and its cataclysm (as recorded by Greek Myths) to create the new Edenic Myth of the Canaanites.

(It is by some legends recorded that the Fall of the Firmament from Heaven and the name of Satan expelled there until he would redeem his works out of the fires of hell, was that incident which separated Hades from the very base of whatever was the land of the earth, from the terrestrial earth, (which philosophers then called the flat earth and which included to them the exegetical ability to obtain the remainder of the solar system and the sun), by the inclusion of the ways within the Labyrinth of whatever was the Firmament tied to Hades; but there in Hades, being Chaos within, the matter was dispensed and never sought out again, except by miracles and without eternity to show the way; except that the Ages must suffer the times of the planets in their passing from one ephemeral domain to another without their knowledge. This in itself was a Judgment and a Fall from the prior Grace to which every high epic age, tried to return in fact – to Eternity to locate the physical domain of an ephemeral truth in the very physical universe of our present abiding.)

But returning again to the problem of the construction and architecture of the astral universe and the terrestrial universe as it is destroyed by carnal warfare against the spirit controlling, we continue to try to establish the place of Eden thereby and that it has no decent or logical conclusion by the very story of itself as well, that thereby, we must be forced to accept. So I ask again and lastly (but almost), "Could it be, Eden was merely a better Canaan?"

Anyway, since the time of any co-linear writ or record, we are to understand that bestiality was given to pastorality by the process of reason together meddled with faith and that this standard alone in creationism was the one in which Mankind found himself at peace; so pastoral Canaan may have been Eden; whereas we hope to find ourselves in the Paradise of Primordial Ages confuting the bestiality of Satanisms by the terms of super-human philosophies.

The confounding which is an ordinary talent to any human imagination concerning the possibility to determine some truth in the physical sense about Eden, does not defy (the human imagination does not ordinarily defy), the point that the Creation Myth of Eden is necessary to preclude the end of self-destruction, by the limited means of objectifying a cataclysmic history in this mythology and   to bring this mythology into the subjection of the judgment of all ages with also a new age attributed to itself; and this then to me is the entire vehicle and purpose of the mythology. That is remains as controlling to all of our scientific and also magical conceptions of philosophical wit and metaphysics especially, is what is so trying to many in the fellowship of literary disciplines to determine a rationale for its story which fits in with the other requirements of faith and reason. But it does not, as we have shown an example today.

This is my interpretation. I shall attempt some silence now on the matter of Eden's realism to the point of a new topic, which I hope proceeds soon. Meanwhile, I could only hope that, these being my opening impressions from merely a perusal of the greater contents without a great study and having read some disquisitions of magical studies in other places, that this theological and demanding puritanical education for and against Satanism is necessary to approaching a clean doctrine of practice in the given discipline.

I would only imagine, however, as I have said herein, that it must be appropriate instruction for us, since we sometimes are led automatically to tell ourselves without any examination that the involvement of our faith and reason to any magical precepts and education, will automatically bring us face to face, not with any spirits of holiness, but with the leadership of holy fallen angels (there being the unholy to account in this; but all the same). Whether true or untrue, we remain aware of the Infernal Furnace and its consequences in these even light-hearted inquisitions on the scantily known topic to our reason and science today. So we forge on now, well-armed with some statements about the origin of magic in the fall of man and that it was a parapet to destroy the magical concept of faith from the hand of the Lord of it. Or so I believe we are saying in this treatise, both the text of origin and also my glosses, though diversely. As principally a theologian and not at all a magician, I feel that the topic of persuasion of religious self-defense beginning with the Tree of Eden, is rather drastic tactic. So I have indemnified it for us. Amen.)

Of which Monsters before mentioned, I will here give a striking example from Helmont.

For he says, a merchant of Aegina, a countryman of his, sailing various times unto the Canaries, was asked by Helmont for his serious judgment about certain creatures, which the mariners frequently brought home from the mountains, as often as they went and called them Tude-squils; for they were dried dead carcasses, almost three-footed and so small that a boy might easily carry one of them upon the palm of his hand.

And they were of an exact human shape, but their whole dead carcass was clear or transparent as any parchment and their bones flexible like gristles; against the Sun, also, their bowels and intestine were plainly to be seen; which thing I, by Spaniards there born, knew to be true.

I considered that, to this day, the destroyed race of the Pygmies were there; for the Almighty would render the expectations of the Evil Spirit, supported by the abominable actions of Mankind, void and vain. And he has, therefore, manifoldly saved us from the craft and subtlety of the Devil, unto whom Eternal Punishments are due, to his extreme and perpetual confusion, unto the Everlasting Sanctifying of the Divine Name.

Chapter 2
Of the Wonders of Natural Magic displayed in a Variety of Sympathetic and Occult Operations throughout the Families of Animals, Plants, Metals and Stones, treated of Miscellaneously.

(For this chapter a word should be slightly defined. More of it may be found in an encyclopedia.

    Archeus. "In alchemy, Archeus, or Archaeus, is a term used generally to refer to the lowest and most dense aspect of the astral plane which presides over the growth and continuation of all living beings."

    This Archeus is what we need to rely on in order to effect what the magnetic powers which the "lower creatures" may have and hold in the fictions of alchemy, (fictions to ourselves, but real of their own rules); and that they may also be true to the experience of the practice of magic; despite that then we learn to understand that we are ourselves these "lower creatures" and not some fabulous creation of spirits which hold the powers of magic dispensable in alchemy. Good enough to the exploration of what is above. Since also, in this treatise, lower creatures are those lower to the ephemeral and appointed angels of heaven and God's spirituals.)

    The Wonders of Animal Magic we mean fully to display under the title of magnetism.

    But here we hasten to investigate by what means, instruments and effects we must apply, actives to passives, to the producing of rare and uncommon effects; whether by actions, amulets, allegations and suspensions – or rings, papers, unctions, suffumigations, allurements, sorceries, enchantments, images, lights, sounds or the like.

    Therefore, to begin with things more simple.

If anyone shall, with an entire new knife, cut asunder a lemon, using words expressive of hatred, contumely or dislike, against any individual, the absent party, though at an unlimited distance, feels a certain inexpressible and cutting anguish of the Heart, together with a cold chilliness and failure throughout the body. Likewise of living Animals, if a live Pigeon be cut through the Heart, it causes the Heart of the party intended to affect with a sudden failure. Likewise, fear is induced by suspending the magical image of Man by a single thread. Also, Death and destruction by means similar to these and all these, from a fatal and magical sympathy.

Likewise of the Virtues of simple Animals, as well as manual operations, of which we shall speak more anon.

The application of a hare's fat, pulls out a thorn. Likewise anyone may cure the toothache with the stone that is in the head of the Toad. Also, if anyone shall catch a living Frog before Sunrise and he or she spits in the mouth of the Frog, (that person) will be cured of an asthmatic consumption. Likewise the right of left eye of the same Animal cures blindness and the fat of a Viper cures a bite of the same. Black Hellebore eases the headache, being applied to the head or (by) the powder snuffed lip (to) the nose in a moderate quantity.

Coral is a well-known preservative against Witchcraft and Poisons, which if worn now, in this time, as much round children's necks as usual, would enable them to combat many diseases which their tender years are subjected to and to which, with fascinations, they often fall a victim. I know how to compose Coral Amulets or Talismans, which, if suspended even by a thread, shall (God assisting) prevent all harms and accidents of violence from Fire or Water or Witchcraft and help them to withstand all their diseases.

Paracelsus and Helmont both agree, that in the Toad, although so irreverent to the sight of man, and so noxious to the touch and of such strong violent antipathy to the blood of Man; I say, out of this hatred, Divine Providence has prepared us a remedy against manifold diseases to man and this scaled image, or idea of hatred, he carries in his head, eyes and most powerfully throughout his whole body.

Now that the Toad may be highly prepared for a sympathetic remedy against the Plague or other disorders, such as the ague, falling sicknesses and various others; and that the terror of us and natural inbred hatred, may the more strongly be imprinted and higher ascend in the Toad, we must hang him up aloft in a chimney, by the legs and set under him a dish of yellow wax, to receive whatsoever may come down or fall from his mouth. Let him hang in this position, in our sight, for three or four days, at least until he is dead. Now we must not omit frequently to be present in sight of the Animal, so that his fears and inbred terror of us, with the ideas of strong hatred, may increase even unto Death.

So you have a most powerful remedy in this one Toad, for the curing of 40,000 persons infected with the pest or plague.

Van Helmont's process for making a preservative amulet against the plague is as follows ~

"In the month of July, in the decrease of the Moon, I took old Toads, whose eyes abounded with white worms, hanging forth into black heads, so that both his eyes were totally formed with worms, perhaps 50 in number, thickly compacted together, their heads hanging out; and as often as anyone of them attempted to get out, the Toad, by applying his forefoot, forbade its utterance. These Toads, being hung up and made to vomit in the manner before mentioned, I reduced the insects and other matters ejected from the Toad, with the waxen dish being added thereto; and the dried carcass of the Toad, being reduced into powder, I formed the whole into troches, with gum-dragon; which being borne about the left breast, drove speedily away all contagion. And being fast bound to the place affected, thoroughly drew out the poison. And these troches were more potent after they had returned into use diverse times than when new.

"I found them to be a most powerful Amulet against the Plague, for if the Serpent eats dust all the days of his Life, because he was the instrument of sinning, so the Toad eats Earth (which he vomits up) all the days of his Life; and, according to the Adept Philosophy, the Toad bears a hatred to man, so that he infects some herbs that are useful to man with his poison, in order for his Death.

"But this difference of note between the Toad and the Serpent ~ the Toad, at the sight of Man, from a natural quality sealed in him, called antipathy, conceives a great terror or astonishment, which terror from Man imprints on this Animal a natural efficacy against the Images of the affrighted Archeus in Man. For truly, the terror of the Toad kills and annihilates the ideas of the affrighted Archeus in Man, because the terror in the Toad is natural (and) therefore radical."

For the poison of the plague is subdued by the poison of the Toad, not by an action primarily destructive, but by a secondary action; as the pestilent idea of hatred or terror extinguishes the ferment, by whose mediation the poison of the plague subsists and proceeds to infect. For seeing the poison of the plague is the product of the image of the terrified Archeus established in a fermental, putrefied odor, and mumial air, this coupling ferments the appropriate mean and immediately the subject of the poison is taken away.

Therefore, the opposition of the Amulet formed from the body, etcetera, of the Toad, takes away and prevents the baneful and most horrible effects of the pestilential poison and ferment of the plague. Hence it is conjectured that he is an Animal ordained by God, that the idea of his terror being poisonous indeed to himself, should be to us, and to our plagues, a poison in terror. Since, therefore, the Toad is most fearful at the beholding of Man, which in himself, notwithstanding, forms the terror conceived from Man and also the hatred against Man, into an Image and active real Being and not consisting only in a confused apprehension; hence it happens that a poison arises in the Toad, which kills the pestilent poison of terror in Man; to wit, from whence the Archeus waxes strong, he not only perceiving the pestilent idea to be extinguished in himself, but, moreover, because he knows that something inferior to himself is terrified, dismayed and does fly.

Again, so great is the fear of the Toad, that if he is placed directly before you and you do behold with an attentive, intentional furious look, so that he cannot avoid you, for a quarter of an hour, he dies, being fascinated with terror and astonishment.

OF THE SERPENT

Hippocrates, by the use of some parts of this Animal, attained to himself Divine Honors; for therewith he cured pestilence and contagion, consumptions and very many other diseases; for he cleansed the Flesh of a Viper. The utmost part of the tail and head being cut off, he stripped off the skin, casting away the bowels and gall. He reserved of the Intestines only the heart and liver. He drew out all the blood, with the vein running down the backbone. He bruised the Flesh and the aforesaid bowels with the bones and dried them in a warm oven until they could be powdered, which powder he sprinkled on honey. Being clarified and boiled, until he knew that the Fleshes in boiling had cast aside their virtue as well in the broth as in the vapors. He then added to this electuary, the spices of his country to cloak the secret.

But this cure of diseases by the Serpent contains a Great Mystery, viz. that as Death crept in by the Serpent of Old, itself ought to be mitigated by the Death of the Serpent; for Adam, being skillful in the properties of all Beasts, was not ignorant also that the Serpent was more crafty than other living Creatures and that the aforesaid Balsam, the remedy of Death, lay hid in the Serpent; wherefore the Spirit of Darkness could not more falsely deceive our first parents than tinder the guileful Serpent's form. For the foolishly imagined they should escape the Death, so sorely threatened by God, by the Serpent's aid.

Amber is an Amulet: a piece of Red Amber worn about one, is a preservative against poisons and the pestilence. Likewise, a Sapphire Stone is as effectual. Oil of Amber, or Amber dissolved in Pure Spirit of Wine, comforts the womb being disordered. If a suffumigations of it be made with the warts of the shank of a Horse, it will cure many disorders of that region. The liver and gall of an Eel, likewise, being gradually dried and reduced to powder and taken in the quantity of a Filbert-Nut in a glass of warm Wine, causes a speedy and safe delivery to women in labor. The liver of a Serpent likewise effects the same.

Rhubarb, on account of its violent antipathy to Choler, wonderfully purges the same. Music is a well-known specific for curing the bit of a Tarantula or any venomous Spider. Likewise, Water cures the Hydrophobia. Warts are cured by paring off the same or by burying as many pebbles, secretly, as the party has Warts.

The King's Evil may be cured by the Heart of a Toad worn about the neck, first being dried. Hippomanes (a substance in the fore lobe of a mare foal) excites lust by the bare touch, or being suspended on the party. If anyone shall spit in the hand with which he struck or hurt another, so shall the wound be cured. Likewise, if anyone shall draw the halter wherewith a malefactor was slain across the throat of one who has the Quinsey (tonsil abscess) it certainly cures him in 3 days.

Also, the Herb Cinque Foil, being gathered before Sunrise, one Leaf thereof, cures the Ague (Malaria or a Rare Fever) of one day; 3 Leaves, cures the Tertian Ague and 4 Leaves, the Quartan Ague.

Rapeseeds, sown with cursings and imprecations, grows the fairer and thrives. But if with praises, the reverse. The juice of deadly Nightshade, distilled and given in a proportionate quantity, makes the party imagine almost whatever you choose.

The Herb Nip, being heated in the hand and afterwards you hold in your hand the hand of any other party, they shall never quit you, so long as you retain that Herb.

The Herb's Arsemart, Comfrey, Flaxweed, Dragon-Wort, Adder's-Tongue, being steeped in cold water and if for some time being applied on a wound, or ulcer, they grow warm and are buried in a muddy place, cures the wound, or sore, to which they were applied. Again, if anyone pluck the Leaves of Asarabacca, drawing them upwards, they will purge another, who is ignorant of the drawing, by vomit only. But if they are wrested downward to the Earth, they purge by stool.

A Sapphire, or a Stone that is of a Deep Blue Color, if it be rubbed on a Tumor, wherein the plague discovers itself (before the party is too far gone) and by and by it be removed from the sick, the absent Jewel attracts all the poison or contagion therefrom.

And thus much is sufficient to be said concerning Natural occult Virtues, whereof we speak in a mixed and miscellaneous manner coming to more distinct headings anon.

Chapter 3
Of Amulets, Charms and Enchantments

The Instrument of Enchanters is a pure, living, breathing Spirit of the Blood, whereby we bind or attract those things which we desire or delight in. So that, by an earnest intention of the Mind, we take possession of the faculties in a no less potent manner than strong Wines beguile the Reason and Senses of those who drink them. Therefore, to charm, is either to bind with words, in which there is Great Virtue, as the poet sings.

"Words three times she spoke, which caused at will, sweet sleep
Appeased the troubled waves and roaring deep."

Indeed, the Virtue of Man's words are so great, that, when pronounced with a fervent constancy of the Mind, they are able to subvert Nature, to cause Earthquakes, Storms and Tempests.

I have, in the country, by only speaking a few words and used some other things, caused terrible rains and claps of thunder. Almost all charms are impotent without words, because words are the speech of the speaker and the image of the thing signified of spoken of; therefore, whatever wonderful effect is intended, let the same be performed with the addition of words in signification of the WILL or DESIRE of the Operator.

For words are a Kind of Occult Vehicle of the Image conceived or begotten and sent out of the Body, by the Soul; therefore, all the forcible power of the Spirit ought to be breathed out with vehemence and an arduous and intent Desire. And I know how to speak and convey words together, so as they may be carried onward to the hearer at a vast distance, no other body intervening, which thing I have done often. Words are also oftentimes delivered to us, seemingly by others, in our sleep, whereby we seem to talk and converse; but then no vocal conversations are of any effect, except they proceed from spiritual and occult causes.

Such Spirits have often manifested singular things to me, while in sleep, the which, in waking, I have thought naught of until conviction of the Truth taught me credulity in such like matters. In the late change of administration, I knew, at least 5 days before it actually terminated, that it would be as I described to a few of my friends.

These things are not alike manifested to everyone. Only, I believe, to those who have long seriously attended to contemplations of this abstruse nature; but there are those who will say it is not so, merely because they themselves cannot comprehend such things. However, not to lose time, we proceed.

There are various Enchantments which I have proven, relative to common occurrences of life, viz. a kind of binding to that effect which we desire; as to love, or hatred; or to those things we love or against those things we hate; in all which there is a magical sympathy above the power of reasoning. Therefore, those abstruse matters we fell, are convinced of and reflect upon and draw them into our use.

I will here set down, while speaking of these things, a very powerful Amulet for the stopping, immediately, a bloody-flux; for the which (with a faith) I dare lay down my life for the success and entire cure.

An Amulet for Flux of Blood

"In the Blood of Adam, arose Death. In the Blood of Christ, Death is extinguished. In the same Blood of Christ, I command you, oh, Blood, that you stop fluxing!"

(And let the party who pronounces these words hold the other's hands in the recital.)

In this one Godly Superstition there will be found a ready, cheap, easy remedy for that dreadful disorder the Bloody Flux, whereby a poor miserable wretch will reap more real benefit than in a whole shop of an apothecary's drugs.

These 4 Letters are a powerful Charm, or Amulet, against the common Ague. Likewise, let them be written upon a piece of clean and new Vellum, at any time of the day or night and they will be found a speedy and certain cure and much more efficacious than the word Abracadabra; however, as that Ancient Charm is still (amongst some who pretend to cure Agues, etcetera) in some repute, I will here set down the form and manner of its being written; likewise, it must be pronounced, or spoken, in the same order as it is written, with the intent or will of the Operator declared at the same time of making it.

Chapter 4
Of Unctions, Philters, Potions, etcetera
Their Magical Virtues

Unguents, or unctions (potions generally), collyries (another word for a salve), philters (an aphrodisiac drink), etcetera, conveying the Virtues of Things Natural to Our Spirits, do multiply, transform, transfigure and transmute it accordingly; they also transpose those Virtues, which are in them, into it, so that it not only acts upon its own body, but also upon that which is near it and affects that (by visible rays, charms and by touching it) with some agreeable quality like to itself.

For, because Our Spirit is the Pure, Subtle, Lucid, Airy and unctuous Vapor of the Blood, nothing, therefore, is better adapted for collyriums, than the like Vapor, which are more suitable to Our Spirit in substance. For then, by reason of their likeness, they do more stir up, attract and transform the Spirit.

The same Virtue have other ointments and confections.

Hence, by the touch, often plague, sickness, faintings, poisoning and love, is induced, either by the hands or clothes being anointed; and often by kissing, things (having) been held in the mouth, love is likewise excited. Now the sight, as it perceives more purely and clearer than the other senses, seals in us the marks of things more acutely and does, most of all and before all others, agree with our fantastic Spirit, as is apparent in dreams, when things seen do more often present themselves to us than things heard or anything coming under the other senses.

Therefore, when collyriums transform the Visual Spirits, that Spirit easily affects the imagination, which, being affected with diverse species and forms, transmits the same, by the same Spirit, unto the outward sense of sight, by which there is formed in it a perception of such species and forms, in that manner, as if it were moved by external objects, that there appear to be seen terrible Images, Spirits and the like.

There are some collyriums which make us see the Images of Spirits of the Air, or elsewhere; which I can make of the Gall of a Man and the Eyes of a Black Cat and some other things. The same is made, likewise, of the Blood of a Lapwing, Bat and a Goat; and if a smooth, shining piece of Steel be smeared over with the Juice of Mugwort, and be made to fume, it causes invocated Spirits to appear.

There are some perfumes or suffumigations and unctions, which make men speak in their sleep, walk and do those things that are done by men that are awake and often what, when awake, they cannot, or dare not do; others, again, make men hear horrid or delightful sounds, noises and the like.

And, in some measure, this is the cause why mad and melancholy men believe they hear and see things equally false and improbable, falling into most gross and pitiful delusions, fearing where no fear is and angry where there is none to contend. Such passions as these we can induce by Magical Vapors, Confections, Perfumes, Collyries, Ungents, Potions, Poisons, Lamps, Lights, etcetera; likewise, by Mirrors, Images, Enchantments, Charms, Sounds and Music; also by diverse Rites, Observations, Ceremonies, Religion, etcetera.

(It's interesting to have to notice that to Alchemy, Religion is another Vehicle or Method or Technique or Material for the execution of magical means. The ideal of metaphysics is not differentiable from the material medium upon which it is exerted. And this fact would be of great benefit to faith, whereas, the mystical ideal must always be without an deserving proposition of effectiveness in mortality and so therefore, it is only some evil magical devise which will be meaningful to those who practice a moveable faith of determined principles.)

MUGWORT
Artemisia vulgaris

©the herbal resource

Chapter 5
Of Magical Suspensions and Allegations ~
showing how and by what Power, they receive Virtue and are efficacious in Natural Magic

When the Soul of the World, by its Virtue, does make all things (that are naturally generated, or artificially made) fruitful, by sealing and impressing on them Celestial Virtues for the working of some wonderful effect, then things themselves not only applied by colliery or suffume, or ointment, or any other such like way; but when they are conveniently bound to, or wrapped up, or suspended about the neck, or any other way applied, although by ever so easy a contact, they do impress their Virtue upon us; by these allegations, etcetera, therefore, the accidents of the body and mind are changed into sickness or health, valor, fear, sadness or joy and the like.

They render those that carry them, gracious, terrible, acceptable, rejected, honored, beloved or hateful and abominable. Now these kind of passions are conceived to be infused no otherwise than in manifest in the grafting of trees, where the vital Life and Virtue is communicated from the trunk to the twig engrafted into it, by way of contact and allegation; so in the female Palm Tree, when she comes near to the male, he boughs bend to the male, which the gardener seeing, he binds them together by ropes across, but soon becomes straight, as if by the continuation of the rope she had received a propagating Virtue from the male.

And it is said, if a woman takes a needle and betray it to dung and put it up into the Earth in which the carcass of a Man has been buried and (thereafter) carry it about her in a piece of cloth used at a funeral, no Man can defile her as long as she carries that.

Now by these examples we see how, by certain allegations of certain things, also suspensions, or by the most simple contact or continuation of any thread, we may be able to receive some Virtues thereby. But it is necessary to know the certain rule of Magical Allegation and Suspension.

And the manner that the Arts requires is this, viz. that they must be done under a certain and suitable Constellation; and they must be done with Wire, or Silken Threads, or Sinews of Certain Animals and those things that are to be wrapped up, are to be done in the Leaves of Herbs, or Skins of Animals, or Membranous Parchments, etcetera.

For if you would procure the Solary Virtue (Spiritual) of anything, this is to be wrapped up in Bay Leaves of the Skin of a Lion, hung around the neck with Gold, Silk or Purple or Yellow Thread. While the Sun reigns in the Heavens, so shall you be endued with the Virtue of that thing.

So if a Saturnine quality or thing be desired, you shall in like manner take that thing, while Saturn reigns and wrap it up in the Skin of an Ass, or in a Cloth used at a Funeral, especially if Melancholy or sadness is to be induced and with a sad or Ash or Leaden or Black Silk or Thread, hang it about your neck and so in the same manner, we must proceed with the rest.

(This chapter distinguishes the alchemies of ancient humors according to a process of material definition in the origin of the natural universe, so that the ether of the magnetic universe is that thing which metaphysically and physically transcends the order of the moment into a universal truth. Or such I understand these theories to mean. However much literalism there is in the ideals remain entirely unknown to myself. So remain magic and not physic to me.)

Chapter 6
Of Antipathies

It is necessary in this place, to speak of the Antipathies of Natural Things, seeing it is requisite, as we go on, to have a thorough Knowledge of that Obstinate Contrariety of Nature, where anything shuns its contrary and drives it, as it were, out of its presence.

Such Antipathy as this has ~

- *The Root Rhubarb against Cholera.*
- *Treacle against Poison.*
- *The Sapphire Stone against Hot Biles, Feverish Heats and Diseases of the Eyes.*
- *The Amethyst against Drunkenness.*
- *The Jasper against the Bloody-Flux and Offensive Imaginations.*
- *The Emerald and Agnus Castus against Lust.*
- *Achates or Agates against Poison.*

(This much is true of medicine today, when particular agates are treated in barium foaming parted chemicals of choice and retreated from a cold, innocuous state to new mutable state in order to be caustic or basal in durable poison from the compulsion of a foreign natural conclusion.)

- *Piony against the Falling Sickness.*
- *Coral against the Ebullition of Black Choler and Pains of the Stomach.*
- *The Topaz against Spiritual Heats.*

(Perhaps we should try our climate controls with Topaz? But no, this is also passion.)

- *Such as are Covetousness, Lust and all manner of Love Excesses.*

(Which is not the problem of the Magma; but otherwise, and for the hope of it, some mystical heat in the Earth. Let's move on.)

The same Antipathy is there, also, of Pismires against the Herb Organ and the Wing of a Bat and the Heart of a Lapwing, from the presence of which they fly. Also, the Organ is contrary to a certain poisonous fly which cannot resist the Sun and resists Salamanders and loathes Cabbage with such a deadly hatred that they cannot endure each other. *So they say Cucumbers hate oil.* (But this could mean for the purpose of plausibility, that Cucumbers are by character are unable to remain without effect in oil; they are mutable to oil and so we interpret hatred; whereas we agree with a love for acerbic acid by the same votive host because the vinegar will coagulate its creature status and begin bestial changes by the former substance.)

And the Gall of a Crow makes even men fearful and drives them from the place wherein it is placed. A DIAMOND disagrees with a LOADSTONE, that, being present, it suffers no Iron to be drawn to it.

Sheep avoid Frog-Parsley as a deadly things and what is more wonderful, Nature has depicted the Sign of this Antipathy upon the Livers of Sheep, in which the very figure of Frog-Parsley does naturally appear.

(There is no apparent document of an herb named Frog Parsley, however there are several varieties of a Parsley Frog, a kind of Frog with more geographic varieties in their architecture. And apparently Sheep despise these particular creatures. But if it refers to an herb, this much is lost, whereas, the other herbs are not.)

Again, Goats hate Garden Basil, as if there was nothing more pernicious. And, amongst Animals, Mice and Weasels disagree, so a Lizard is of a contrary nature to a Scorpion and induces great terror to the Scorpion with its very sight and they are therefore killed with the oil of them.

(This is also an interesting mythological antiquity in which the Basilisk and the Scorpion are enemies.)

There is a great enmity between Scorpions and Mice; therefore if a Mouse be applied to the bite of a Scorpion, he cures it. Nothing is so much an enemy to Snakes as Crabs; and if Swine be hurt by them (Snakes), they are cured by Crabs. The Sun, also being in Cancer, Serpents are tormented. Also, the Scorpion and Crocodile kill one another.

(The Scorpion has no natural approval besides some Meerkat creature which will eat them, but to whom the sting is nevertheless deadly, so we should submit them for friends.)

And if the Bird Ibis does but touch a Crocodile with one of his Feathers, he makes him immoveable. The Bird called a Bustard (a Buzzard) flies away at the sight of a Horse and a Hart at the sight of a Ram, or a Viper. An Elephant trembles at hearing the grunting of a Hog, so does a Lion at the crowing of a Cock. And a Panther will not touch them that are anointed with the fat of a Hen, especially if Garlic has been put to it.

There is also an enmity between Foxes and Swans, Bulls and Jackdaws (Grey Crow). And some Birds are at a perpetual variance, as Daws and Owls, Kites and Crows, Turtle and Ring-Tail, Egepis and Eagles; also, Harts and Dragons (large land or amphibious four legged Reptiles).

Amongst Water Animals, there is a great Antipathy between Dolphins and Whirlpools, the Mullet and Pike, Lamprey and Conger, Pour-Contrel and Lobster, which the latter, but (in) seeing the former, is nearly struck dead with fear, but the Lobster tears the Conger.

(A Pour-Contrel is in the Fish category of Octopus and Squid somewhere and a Conger is in the category of Congrid Eels and large.)

The Civet-Cat cannot resist the Panther and if the Skins of both be hung up against each other, the Skin or Hairs of the Panther will fall off.

Apollo says, in his Hieroglyphics, if anyone be girt about with the Skin of a Civet-Cat, he may pass safely through his enemies.

The Lamb flies from the Wolf and if the Tail, Skin or Head of Lupus be hung up in the Sheep's-Cot, they cannot eat their meat for very fear.

And Pliny mentions the Bird called the Marlin, that breaks the Eggs of the Crow, whose young are annoyed by the Fox, that she also will pinch the Whelps of the Fox and the Fox likewise, which, when the Crow sees, they help the Fox against her as against a common enemy.

The Linnet lives in and cats Thistles, yet she hates the Ass, because he cats Thistles and flowers them. (A Linnet is a brown and grey, red-breasted Finch, but it cultivates flowers to thistles.)

There is so great an enmity between the Little Bird called Esalon and the Ass, that their Blood will not mix and that, at the simply braying of the Ass, both the Esalon's Eggs and young, perish together.

(The Esalon has no fellowship except in ancient literature of magic instruction to an easy record, so I image they might mean, also having read about their magical importance, a Tit, a Titmouse, or a Chickadee, or any of those that might satisfy the same.)

There is also a total Antipathy of the Olive Tree to the Harlot. That, if she plant it, it will neither thrive nor prosper, but wither. (Beware to the Satyr kind for insubstantial survival.)

A Lion fears lighted torches and is tamed by nothing sooner. The Wolf fears not sword or spear, but a Stone, by the throwing of which, a wound being made, Worms breed in the Wolf.

A Horse fears a Camel so much that he cannot endure the picture of the Beast. An Elephant, when he rages, is quieted by seeing a Cock. A Snake is afraid of a Naked Man, but pursues one clothed.

A Mad Bull is tamed by being tied to a Fig Tree.

Amber attracts all things to it but Garden Basil and things smeared with oil, between which there is Natural Antipathy.

(So it is not an effective potent for relationships, as it is not.)

Chapter 7

Of the OCCULT VIRTUES ~
(and) of Things which are Inherent in them only in their Lifetime; and such as remain in them even after Death.

    It is expedient for us to know that there are somethings which retain Virtue only while they are living, others even after Death. So in the Colic, if a live Duck be applied to the Belly, it takes away the pain and the Duke dies.
    If you take the Heart out of any Animal, and, while it is warm, bind it to one that has a Quartan Fever, it drives it away. So if anyone shall swallow the Heart of a Lapwing, Swallow, Weasel, or a Mole, while it is yet living and warm with Natural Heat, it improves his intellect and helps him to remember, understand and foretell things to come. Hence this general rule ~ that whatever things are taken for Magical Uses from Animals, whether they are stones, members, hair, excrements, nails, or anything else, they must be taken from those Animals while they are yet alive, and, if it is possible, that they may live afterwards. If you take the Tongue of a Frog, you put the Frog into Water again.
    And Democritus writes, that if anyone shall take out the Tongue of a Water Frog, no other part of the Animal sticking to it and lay it upon the place where the Heart beats of a Woman, she is compelled, against her will, to answer whatsoever you shall ask of her.
    Also, take the Eyes of a Frog, which must be extracted before Sunrise and bound to the sick party and the Frog to be let go again blind into the Water, the party shall be cured of a Tertian Ague; also, the same will, being bound with the Flesh of a Nightingale in the Skin of a Hart, keep a person always wakeful without sleeping.

Also, the Roe of the Fork Fish, being bound to the Navel, is said to cause women an easy childbirth, if it be taken from it alive and the Fish put into the Sea again. So the Right Eye of a Serpent, being applied to the soreness of the Eyes, cures the same, if the Serpent be let go alive. So, likewise, the Tooth of a Mole, being taken out alive, and afterwards let go, cures the Toothache and Dogs will never bark at those who have the Tail of a Weasel that has escaped.

Democritus says, that if the Tongue of the Chameleon be taken alive, it conduces to good success in trials and likewise to Women in labor; but it must be hung up on some part of the outside of the House, otherwise, if brought into the House, it might be most dangerous.

There are very many properties that remain after Death and these are things in which the Idea of the Matter is less swallowed up, according to Plato, in them: even after Death, that which is Immortal in them will work some wonderful things. As in the Skins we have mentioned of several Wild Beasts, which will corrode and eat one another after Death; also, a drum made of the Rocket Fish, drives away all creeping things at what distance soever the sound of it is heard. And the Strings of an Instrument made of the Guts of a Wolf and being strained upon a Harp or Lute, with Strings made of Sheep Guts, will make no Harmony.

Chapter 8
Of the Wonderful Virtues of Some Kind of Precious Stones

It is a common opinion of Magicians, that Stones inherit Great Virtues, which they receive through the Spheres and Activity of the Celestial Influences, by the Medium of the Soul or Spirit of the World.

Authors very much disagree in respect of the probability of their actually having such VIRTUES in POTENTIA, some debating warmly against any OCCULT or SECRET VIRTUE lying hid in them. Others (will show) as warmly, the causes and effects of these sympathetic properties.

However, to leave these trifling arguments to those who love cavil and contentions better than I do, and, as I have neither leisure nor inclination to enter the lists with sophists, and tongue-philosophers, I say, that these Occult Virtues are disposed throughout the Animal, Vegetable and Mineral Kingdoms, by seeds or Ideas, originally emanating from the Divine Mind and through Super-Celestial Spirits and Intelligence always operating, according to their proper offices and governments allotted them. Which Virtues are infused, as we before said, through the Medium of the Universal Spirit, as by a general and manifest sympathy and antipathy established in the Law of Nature.

Amongst a variety of examples, the Loadstone is one most remarkable proof of the sympathy and antipathy we speak of. However, to hasten to the point. Amongst Stones, those which resemble the Rays of the Sun by their Golden Sparklings (as does the glittering Stone Aetites) prevent the Failing Sickness and Poisons, if worn on the Finger, -- so the Stone which is called Oculis Solis, or Eye of the Sun, being in figure like to the Apple of the Eye, from which shines forth a ray, comforts the Brain and strengthens Sight.

(A Loadstone, or Lodestone, is one that can has magnetic properties and allows itself to be used in the construction of the same.)

The Carbuncle, which shines by Night, has a Virtue against all airy and vaporous Poisons. The Chrysotile Stone, of a light Green color, when held against the Sun, there shines in it a Ray like a Star of Gold. This is singularly good for the Lungs and cures Asthmatical Complaints. And if it be bored through and the hollow filled with the Mane of an Ass and bound to the Left Arm, it chases away all Foolish and Idle Imaginations and Melancholy Fears and drives away Folly.

The Stone called Iris, which is like Crystal in Color, being found with Six Corners, when held in the Shade and the Sun suffered to shine through it, represents a Natural Rainbow in the Air.

The Stone, Heliotropium, Green, like a Jasper or Emerald, beset with Red Specks, makes the wearer constant, renowned and famous and conduces to long Life. There is, likewise, another wonderful property in this Stone and that is, that it so dazzles the Eyes of Men, that it causes the bearer to be Invisible, but then there must be applied to it the Herb bearing the same name, viz. "heliotropium," or the Sunflower. And these Kind of Virtues, Albertus Magus and William of Paris, mention in their writings.

The Jacinth also possesses Virtue from the Sun against Poisons, Pestilences and Pestiferous Vapors; likewise it renders the bearer pleasant and acceptable; conduces also, to gain money; being simply held in the mouth, it wonderfully cheers the Heart and strengthens the Mind. Then there is the Pyrophilus, of a Red Mixture, which Albertus Magnus reports that Aesculapius makes mention of in one of his Epistles to Octavius Caesar, saying.

"There is a certain Poison, so intensely Cold, which preserves the Heart of Man, being taken out, from burning; so that if it be put into the Fire for anytime, it is turned into a Stone, which Stone is called Pyrophilus."

It possess a wonderful virtue against Poison and it infallibly renders the wearer thereof renowned and dreadful to his enemies.

Apollonius is reported to have found a Stone called Pantaura (which will attract other Stones, as the Loadstone does Iron) most powerful against all Poisons. It is spotted like the Panther and therefore some Naturalists have given this Stone the name of Pantherus. Aaron calls it Evanthum. And some, on account if its variety, call it Pantochras.

Chapter 9

Of the Mixtures of NATURAL THINGS one with another; and the Production of MONSTROUS ANIMALS by the Application of NATURAL MAGIC

Magicians, Students and Observers of the Operations of Nature, know how, by the application of active forms to a Matter fitly disposed and made, as it were, a proper recipient to effect many wonderful and uncommon things that seem strange and above Nature, by gathering this and that thing beneficial and conducive to that effect which we desire. However, it is evident that all the Powers and Virtues of the Inferior Bodies are not found comprehended in any one single thing, but are dispersed amongst many of the compounds here amongst us. Wherefore, it is necessary, if there 100 Virtues of the Sun dispersed through so many Animals, Plants, Metals or Stones, we should gather all these together and bring them all into one form, in which we shall see all the said Virtues, being united, contained. Now there is a Double Virtue in commixing; one, viz. which was once planted in its parts and is CELESTIAL; the other is obtained by a certain Artificial Mixture of Things, mixed among themselves, according to a due proportion, such as agree with the Heavens under a certain Constellation. And this virtue descends by a certain similitude or likeness that is in things amongst themselves, by which they are drawn or attracted towards their superiors and as much as the following do by degrees correspond with them that go before, where the patient is fitly applied to its agent.

So from a certain Composition of Herbs, Vapors and such like, made according to the Rule of Natural and Celestial magic, there results a certain common form; of which we shall deliver the true and infallible Rules and Experiments in our Second Book, where we have written expressly on the same.

We ought, likewise, to understand that by how much more Noble and Excellent the form of anything is, by so much the more it is prone and apt to receive and powerful to act. Then the Virtue of Things do indeed become Wonderful. Viz. when they are applied to Matters, mixed and prepared in fit seasons to give them life, by procuring life for them from the Stars, our own Spirit powerfully co-operating therewith. For there is so great a power in prepared Matters, which we see do then receive Life, when a perfect mixture of qualities do break the former contrariety; for so much the more perfect Life things receive, by as much the temper and composition is free from contrariety.

Now the Heavens, as a prevailing cause, do, from the Beginning of Everything (to be generated by the concoction and perfect digestion of the Matter), together with Life, bestow Celestial Influences and Wonderful Gifts, according to the capacity that is in that Life and Sensible Soul to receive more Noble and Sublime Virtues. For the Celestial Virtue otherwise lies asleep, as Sulphur kept from Flame. But in Living Bodies it does always burn, as kindled Sulphur, which, by its Vapor, fills all the places that are near.

There is a book called, "A Book of the Laws of Pluto," which speaks of Monstrous Generations, which are not produced according to the Laws of Nature. Of these things which follow we know to be true; viz. of Worms are generated Gnats; of a Horse, Wasps; of a Calf and Ox, Bees. Take a Living Crab, his Legs being broken off and he buried under the Earth, a Scorpion is produced. If a Duck be dried into powder and put into Water, Frogs are soon generated; but if he be baked in a pie and cut into pieces and be put in a moist place underground, Toads are generated.

Of the Herb Garden Basil, bruised and put between Two Stones, are generated Scorpions. Of the Hairs of Menstruous Woman, put under Dung, are bred Serpents; and the Hair of a Horse's Tail, put into Water, receives Life and is turned into a most pernicious Worm. And there is an Art wherewith a Hen, sitting upon Egg, may be generated the Form of a Man, which I myself know how to do and which Magicians call the Mandrake and it has in it Wonderful Virtues.

You must, therefore, know which and what kind of Matters are either of Art of nature, begun or perfected or compounded of more things and what Celestial Influences they are able to receive. For a congruity of Natural Things is sufficient for the receiving of influence from Celestial, because, nothing hindering, the Celestials send forth their light upon inferiors they suffer no matter to be destitute of their Virtue. Wherefore, as much Matter as is perfect and pure is, as we before said, fitted to receive Celestial Influences; for that is the binding and continuing of the Matter of the Soul to the World, which does daily flow in upon things Natural and all things which Nature has prepared, that is impossible that a prepared Matter should not receive Life or a more Noble Form.

Chapter 10
Of the Art of Fascination, Binding, Sorceries, Magical Confections, Lights, Candles, Lamps, etcetera, etcetera; ~
Being the Conclusion of the NATURAL MAGIC

We have so far spoken concerning the Great Virtues and wonderful efficacy of Natural Things.

It remains now that we speak of a Wonderful Power and Faculty of Fascination. Or, more properly, a Magical and occult binding of men into Love or Hatred, Sickness or Health.

Also, the binding of thieves, that they cannot steal in any place, or to bind them that they cannot remove, from whence they may be detected;

- the binding merchants, that they cannot buy or sell;
- the binding of an army, that they cannot pass over any bounds;
- the binding of ships, so that no wind, though ever so strong, shall be able to carry them out of that harbor;
- the binding of a mill, that it cannot, by any means whatsoever, be turned to work;
- the binding of a cistern, or fountain, that the water cannot be drawn up out of them;
- the binding of the ground, so that nothing will bring forth fruit or flourish in it, also, that nothing can be built upon it
- the binding of fire, that, though it be ever so strong, it shall burn no combustible thing that is put to it; also,
- the binding of lightnings and tempests, that they shall do no hurt
- the binding of dogs, that they cannot bark; also,
- the binding of birds and wild beasts, that they shall not be able to run of fly away
- and things similar to these, which are hardly credible, yet known by experience.

Now how it is that these kind of buildings are made and brought to pass, we must know. They are thus done; by sorceries, collyries, unguents, potions, binding to and hanging up of Talismans, by charms, incantations, strong imaginations, affections, passions, images, characters, enchantments, imprecations, lights and by sounds, numbers, words, names, invocations, swearings, conjurations, consecrations and the like.

OF SORCERIES

The Force of Sorceries are, no doubt, very powerful; indeed, they are able to confound, subvert, consume and change all inferior things.

Likewise, there are Sorceries by which we can suspend the Faculties of Men and Beasts.

Now, a we have promised, we will show what some of these Kind of Sorceries are, that, by the example of these, there may be a way opened for the whole subject of them.

Of these, the First is Menstruous Blood; which, how much power it has in Sorcery, we will now consider.

First, if it comes over new Wine, it will turn it sour. And if it does but touch a Vine, it will spoil it forever. And, by its very touch, it renders all Plants and Trees Barren and those newly set, die. It makes dim the brightness of a looking-glass, dulls the edges of knives and razors, dims the beauty of polished ivory and makes iron rusty.

It likewise, makes Brass rusty and to smell very strong. By the taste, it makes Dogs run mad, and, being thus mad, if they once bite anyone, that wound is incurable.

It destroys whole hives of bees and drives them away, if it does but touch them. It makes linen black that is boiled with it. It makes Mares cast their Foals by touching them with it and Women miscarry. It makes Asses barren if they cat of the Corn touched by it.

The Ashes of Menstruous Clothes cast upon Purple Garments, that are to be washed, change their color and likewise take away the color of flowers. It also drives away Tertian and Quartan Agues, if it be put into the Wool of a Black Ram and tied up in a Silver Bracelet.

As also if the Soles of the Patient's Feet be anointed therewith and especially if it be done by the Woman herself, the Patient not knowing what she uses. It likewise cures the Falling Sickness. But most especially, it cures them that are afraid of Water or Drink after they are bitten by a mad Dog, if only a monstrous cloth be put under the cup.

Likewise, if a Menstruous Woman shall walk naked before Sunrise, in a Field of Standing Corn, all hurtful things Perish. But, if after Sunrise, the Corn withers.

Also, they are able to expel hail, rain, thunders and lightnings; more of which Pliny mentions.

Know this, that if they happen at the decrease of the Moon, they are a much greater Poison than in the increase and yet much greater if they happen between the decrease and change. But if they happen in the Eclipse of the Sun or Moon, they are a most incurable and violent Poison. But they are of the greatest force when they happen in the first years of the Virginity, for then if they but touch the doorposts of a house, no mischief can take effect in it.

And some say that the threads of any Garment touched therewith cannot be burnt and if they are cast into a Fire, it will spread no farther. Also it is noted, that the root of Peony, being given with Castor and smeared over with a Menstruous Cloth, it certainly cures the Falling Sickness.

Again, let the Stomach of a Hart be roasted and to it be put a Perfume made with a Menstruous Cloth. It will make Crossbows useless for the killing of any Game.

The Hairs of a Menstruous Woman, put under dung, breeds Serpents. And if they are burnt, will drive away Serpents with the fume. So great and powerful a Poison is in them, that they are a Poison to Poisonous Creatures.

We next come to speak of Hippomanes, which, amongst Sorceries, are not accounted the least.

And this is a little venomous piece of Flesh, the size of a Fig and black, which is in the Forehead of a Colt newly Foaled, which, unless the Mare herself does presently eat, she will hardly ever lover her Foals or let them suck. And this is a most powerful Philter to cause Love, if it be powdered and drank in a cup with the blood of him that is in love.

Such a Potion was given to Medea by Jason.

There is another Sorcery which is called Hippomanes, viz. a venomous Liquor issuing out of the share of a Mare at the time she Lusts after the Horse.

The Civet Cat (African or Asian, a small, larger than ferret sized or ferret sized Lynx, Ocelot, Caracal, Wildcat, Jaguar, Bobcat or kind), also, abounds with Sorceries. For the posts of a door, being touched with her (the Cat's) blood, the Arts of Jugglers and Sorcerers are so invalid that Evil Spirits can by no means be called up or compelled to talk with them.

This is Pliny's report.

Also, those that are anointed with the Oil of her Left Foot, being boiled with the Ashes of the Ankle Bone of the same and the Blood of a Weasel, shall become odious to all. The same, also, is to be done with the eye being decocted.

If anyone has a little of the Strait Gut of this Animal about him and it is bound to the Left Arm, it is a Charm. That if he does but look upon a Woman, it will cause her to follow him at all opportunities; and the Skin of this Animal's Forehead withstands Witchcraft.

We next come to speak of the Blood of a Basilisk, which Magicians call the Blood of Saturn.

This procures (by its Virtue) for him that carries about him, good success of petitions from great men. Likewise makes him amazingly successful in the cure of diseases and the grant of any privilege.

They say, also, that a tyke, if it be taken out of the left car of a dog and it be altogether black, if the sick person shall answer him that brought it in, and who, standing at his feet, shall ask him concerning his disease, there is certain hope of life. And that he shall die if he make him no answer.

They say, also, that a stone bitten by a mad Dog causes discord, if it be put into drinks. And if anyone shall put the tongue of a Dog, dried, into his shoe, or some of the powder, no Dog is able to bark at him who has it. And more powerful this, if the Herb, Hound's Tongue be put with it.

And the Membrane of the Secundine (placenta) of a Bitch does the same. Likewise, Dogs will not bark at him who has the Heart of a Dog in his pocket.

The Red Toad, Pliny says, living in briers and brambles, is full of Sorceries and is capable of Wonderful Things. There is a little Bone in his left side, which, being cast into cold Water, makes it presently hot. By which, also, the rage of Dogs are restrained and their love procured, if it be put in their drink, making them faithful and serviceable. If it be bound to a Woman, it stirs up Lust.

On the contrary, the bone which is on the right side makes hot Water, cold and it binds it so that no heat can make it hot while it there remains. It is a certain cure for Quartans (Ague), if it be bound to the sick in a Snake's Skin. And likewise, cures all Fevers, the St. Anthony's Fire and restrains Love and Lust. And the Spleen and Heart are effectual Antidotes against the Poisons of the said Toad.

Thus much Pliny writes.

(Considering this information comes to us from Pliny for the purpose of Sorcery, we will use it to our topic and the keeping of that it has been given to us.

Rationally, it is unlikely that we are advising ourselves at this present time to use any of the conclusions for magical sources of powers to ourselves or others. The ideal inherent of the writ of this text provides us with some assurance of whatever topics we might be approaching concerning the occult and its use of magic.

That it is indeed Theosophy and that this Theosophy is actually quite rooted in the Grander Knowledge, the Established Knowledge of Doctrina Antiqua.

What it will mean to us in this later Scientific Age, has been little ciphered except by Masters of Theosophies and those topics the same, which include scientific disciplines concerning our present times and of relation to whatever is by science, affected by Occult Science, and especially those Ancient (Occult) Ones, Mysteries, Heresies and Mythologies, considered to be of the choicest and most secret and most important kind of Knowledge and those which are nevertheless today, considered of less value of Knowledge to the former-told exercise of Science, "our own".

However, to consider one's self learned in some discipline of Magic, should consider the responsibility to sit for the first lessons of it and also to assure one's self that the entire domain is not therefore or thereby entertained or entered. But that to begin in a place with an appropriate authority is to understand some valuable mystery to one's own account of Knowledge and whatever is lost or void in it.

And thus it is we have chosen this course today. In the name of Ancient Doctrines and their being lost to us and still leading us down pathological corridors of doubt with our own treatises and theses, wondering if those Ancients had some better insights for us, if we had continued to strive through ruin with them. So we attempt to do with professors of an art and especially the macabre domain of what magic portends to the more modern age in soliloquy of past ages and their uses of it known; that, we do not understand what part of miracles were true and part of falsehoods and catastrophes were false, except that we should become educated in some faith of what we have as Theosophy. We must remain with our topic to begin it for a more sophisticated end later with our own questions of the grander schemes, whatever they are.)

Also it is said, that the Sword with which a Man is slain has wonderful power.

For if the snaffle of a bridle, or bit, or spurs, be made of it, with these, a Horse ever so wild is tamed and made gentle and obedient.

They say, if we dip a Sword, with which anyone was beheaded, in Wine, that it cures the Quartan, the Sick being given to drink of it.

There is a Liquor made, by which men are made as raging and furious as a Bear, imagining themselves in every respect to be changed into one. And this is done by dissolving or boiling the Brains and Heart of that Animal in New Wine and giving anyone to drink out of a skull, and, while the force of the draught operates, he will fancy every Living Creature to be a Bear like to himself.

Neither can anything divert or cure him until the Fumes and Virtue of the Liquor are entirely expended, no other Distemper being perceivable to him.

The most certain cure of a violent Headache, is to take any Herb growing upon the top of the Head of an Image. The same being bound or hung about one with a Red Thread, it will soon allay the violent pain thereof.

OF MAGICAL LIGHTS, CANDLES, LAMPS, etcetera.

There are made, artificially, some kinds of Lamps, Torches, Candles and the like, of some certain and appropriate Materials and Liquors opportunely gathered and collected for this purpose, which, when they are lighted and shine alone, produce some wonderful effects.

There is a Poison from Mares, after copulation, which, being lighted in torches composed of their Fat and Marrow, does represent on the walls, a monstrous deformity of Horses' Heads, which thing is both easy and pleasant to do. The like may be done of Asses and Flies.

And the Skin of a Serpent or Snake, lighted in a Green Lamp, makes the Images of the same to appear. And Grapes produce the same effect, if, when they are in their Flowers, you shall take a Phial and bind it to them, filled with Oil; and shall let them remain so until they are ripe and then the Oil be lighted in a Lamp, you shall see a prodigious quantity of Grapes. And the same in other Fruits.

If Centaury be mixed with Honey and the Blood of a Lapwing and be put in a Lamp, they that stand about will be of a gigantic stature. And if it be lighted in a clear evening, the Stars will seem scattered about. The Ink of the Cuttlefish, being put into a Lamp, makes Blackamoors (Black African Moors, or more accurately their Spirits) appear.

So also, a Candle made of some Saturnine things, such as Man's Fat and marrow, the Fat of a Black Cat, with the Brains of a Crow or Raven, which being extinguished in the Mouth of a Man lately dead, will afterwards, as often as it shines alone, bring great Horror and Fear upon the spectators about it.

Of such like Torches, Candles, Lamps, etcetera (of which we shall speak further in our Book of Magnetism and Mummies), Hermes speaks largely of; also Plato and Chyrannides.

And of the later Writers, Albertus Magnus makes particular mention of the Truth and Efficacy of these, in a Treatise on these particular things relative to Lights, etcetera.

OF THE ART OF FASCINATION OR BINDING BY THE LOOK OR SIGHT

We call Fascination a Binding because it is effected by a Look, Glance or Observation, in which we take possession of the Spirit and over-power the same, of those we mean to fascinate or suspend. For it comes through the eyes and the instrument by which we fascinate or bind is a certain, pure, lucid, subtle Spirit, generated out of the ferment of the purer blood by the heat of the heart and the firm, determined and ardent Will of the Soul which directs it to the object previously disposed to be fascinated. This does always send forth by the eyes, rays or beams, carrying with them a pure, subtle Spirit or Vapor into the eye or blood of him or her that is opposite.

So the Eye, being opened and intent upon anyone with a strong imagination, does dart its beams, which are the Vehicle of the Spirit, into whatever we will affect or bind, which Spirit, striking the Eye of them who are fascinated, being stirred up in the Heart and Soul of him that sends them forth and possessing the Breast of them who are struck, wounds their Hearts, infects their Spirits and over-powers them. Know, likewise, that in Witches, those are most bewitched, who, with often-looking, direct the edge of their sight to the edge of the sight of those who bewitch or fascinate them. Whence arose the saying of Evil Eyes, etcetera. For when their eyes are reciprocally bent one upon the other and are joined beams to beams and lights to lights, then the Spirit of the one is joined to the Spirit of the other and then are strong ligations made. And most violent Love is stirred up, only with a sudden looking on, as it were, with the darting a look or piercing into the very inmost of the Heart, whence the Spirit and amorous Blood, being thus wounded, are carried forth upon the Lover and Enchanter.

Not otherwise than the Spirit and the Blood of him that is murdered is upon the Murderer, who, if standing near the body killed, the Blood flows afresh which thing has been tried by repeated experiments. So great power is there in fascination that many uncommon and wonderful things are thereby effected, especially when the Vapors of the Eyes are subservient to the Affection.

Therefore, collyries, ointments, allegations, etcetera, are used to affect and corroborate the Spirit in this or that manner. To induce Love, they use Venereal Collyriums, as Hippomanes, Blood of Doves, etcetera. To induce Fear, they use Martial Collyriums, as the Eyes of Wolves, Bear's Fat and the Civet-Cat.

(Collyrium is (also, besides as a salve), defined as dark overshadowing put over and upon the eyes for heightening the senses of the same into transcending visions with concerns to spiritual matters.)

To procure Misery or Sickness, they use Saturnine and so on.

Thus much we have thought proper to speak concerning Natural Magic, in which we have, as it may be said, only opened the First Chamber of Nature's Storehouse.

Indeed, we should have inserted many more things here, but as they fall more properly under the headings of Magnetism, Mummy, etcetera, to which we refer the Reader, we shall take our leave of the Reader for the present, that we may give him time to breathe, likewise to digest what he has here feasted upon. And, while he is preparing to enter the unlocked Chambers of Magic and Nature, we will procure him a rich service of most delicious meats, fit for the hungry and thirsty traveler through the vast Labyrinths of Wisdom and True Science.

(But we have alleged like bad students, that this is no True Science, even if it surely speaks of a very Occult Wisdom to our orientation and perspective on the causes and effects of physical and natural sciences. As before in the Doctrina Antiqua, I allege that the occult mysteries of our curiosities still today, have no open door to trespass upon by means of our lacking a full science of astrophysical principles which remain to be "true" insomuch of practice and theory together. So I could understand that the statements of these Occult ideals, because they are only ideals to us of the day's persuasions and legalisms, that they say early on, Barrett does say early on, on the behalf of himself and his professorship in the esteem of the most astoundingly influential Ancient Philosophers, that there is a definition of an astrophysical connection which needs to be overcome as unscientific in order to put to rest the idea of Occult Science from whatever other Science there is. There is no completion of today's astrophysical domain of practice to complete theory with application and so we have no definitive approach without magic to answering too many mysteries to ourselves concerning whatever is Science in the universe and coming down to the Earth from the same venue of scientific research. What is essentially witchcraft remains that to us without any other Science but the Occult and there are more mysteries of myriad mythologies besides the comfort of that title of magic; so the problem is the problem of faith in science and faith in immaterial things, faith in authority which is historically reliable to a point and faith in newer intelligence which is practical to a point, and the drive to understand and find dominion over the unsolved mysteries of occult knowledge. This usually leads down a pathological path of occult science which is astrophysical bent of persuasion to find determinism in that, what I say is an unfinished Science today, by the actual incompletion of terms to applications; whereas, most of mechanical science has revealed itself that way, except in the alter-dimensions of the astral and esoteric planes of space. We need to better understand Magic and so we need to adopt its approach to Science, especially when it is given in the authority of the Ages, which have been benevolent to us for a large part and with friendship in mind.)

THE END OF THE NATURAL MAGIC

The Author, having, under the Title of Natural Magic, collected and arranged everything that was curious, scarce and valuable, as well his own experiments, as those in which he has been indefatigable in gathering from the Science and Practice of Magical Authors, and those the most Ancient and abstruse, as may be seen in the list at the End of the Book, where he has put down the names of the Authors, from which he has translated many things that were never yet published in the English language, particularly Hermes, Tritemius, Paracelsus, Bacon, Dee, Porta, Agrippa, etcetera, etcetera, etcetera; from whom he has not been ashamed to borrow what he thought and knew would be valuable and gratifying to the Sons of Wisdom, in addition to many other rare and uncommon experiments relative to this Art.

PART 2

in Two Sections

1. Of the Divine Origin of Alchemy
2. Of the Matter of the Philosopher's Stone

The MAGUS

THE JEWEL OF ALCHEMY

Or, The True Secret of the Philosopher's Stone

Wherein the process of making the GREAT ELIXIR is discovered; by which BASE METALS may be turned into PURE GOLD;

containing the Most Excellent and Profitable Instructions in the HERMETIC ART;

discovering that valuable and SECRET MEDICINE of the PHILOSOPHORS, to make Men Healthy, Wise and Happy.

By F. Barrett, Student of Chemistry, Natural Philosophy, 1801

EPISTLE TO MUSEUS

Oh Museus! Whose mind is High! Observe my words and read them with your own Eye. These Secrets in your Sacred Breast, re-open and in your Journey think of God alone: the Author of all Things that cannot die, of Whom we now shall speak.

I tell you here, Museus, to observe our Words and read them with your Eye, that is, the Eye of your Understanding, for know, there are many that hear us speak, that read not the meaning of our Words. Wherefore, should you contemplate these Mysteries with so much constancy of Mind, if you did not perceive in them some great good most desirable!

Listen, then, oh young man and hear our Words!

We will show you the dangerous precipice of Vanity and headlong Desire. We will describe to you the stubborn and fatal Will of our Passions, even with Tears of Contrition and heartfelt Compassion for you inexperience. We will lead you, as it were, by the hand, through those Labyrinths of Vice, wherewith you are daily surrounded and, however prejudiced you might be against the receiving of Our Doctrine, yet be assured, we have in our Possession the Magical Virtue and Power of Binding you to Our Principles and making you Happy, inspite of yourself.

Here is a Great Secret! You shall say, every man wishes to be happy, which I grant, but my answer is, most men prevent their own happiness. They destroy it, by suffering themselves to be governed by the outward Principle of the Flesh, thinking the greatest good to be in the satisfying of their Carnal Appetites or in the amassing together heaps of Wealth, whereby they thrust down the meek and poor, raising up the Standards of Pride, Envy and Oppression.

These things every day's experience confirms. Nay, there are some so blind, that, in the possession of much Wealth, they think there is nothing beyond it. Insomuch, that they triumph in Lust, Oppression, Revenge and Contumely. But how is it, you will say, that, seeing Man is a reasonable Being, he can possibly give up his government so easily? I say, when Man suffers the unreasonable and Bestial Part to deprave him, then he immediately becomes a slave (and the vilest of slavery is that which deprives Man of his Social Virtues), for then, although in the possession of great worldly things, such as houses, estates and all other temporal gifts, yet he becomes an immediate instrument to the Prince of this World and the Powers of Darkness, seeing that those riches he inherits are merely given him in this Life, to bestow upon others those necessaries and comforts which he himself does not feel the want of and by which he might, if not blinded by his Passions and Lusts, secure himself an Eternal and Incorruptible Treasure.

But he who possesses Treasures without Mercy, Liberty, bounty, Charity, etcetera, robs the Eternal Author of all Good, of the Honor due unto him, and, in short, is working destruction to his own Soul; his riches, instead of benefitting himself and others, eventually and finally terminates as a curse. While he lives here he is a scourge to society and after he leaves this, it is plain enough pointed out in the New Testament what will he his situation and condition.

Therefore, you young man, that has but a few years to live, study how to attain the Stone we teach of. It will protract the Beauty of your youth, though you should live for centuries; it will ever supply you with the means of comforting the afflicted; insomuch, that when you have attained this truly desirable and most perfect Talisman, your Life will become soft and pleasant. No cares, nor corroding pangs, no self-torment will ever invade your Mind; neither shall you want the means to be happy, in respect of the Possession of the Goods of this Life, but shall have abundantly. But how and from what source all this is to proceed, out of what Thing or Matter, you shall attain your Wished-for-End; the studying of the ensuing Treatise will sufficiently show.

Your Friend, F.B.

TO THE READER

Although we do not, in any point of Science, arrogate perfection in ourselves, yet something we have attained by dear experience, by diligent labor and by study, worthy of being communicated for the instruction of either the licentious Libertine or the grave Student – the Observer of Nature. And this, our Work, we concentrated into a Focus. It is, as it were, a Spiritual Essence drawn from a large quantity of Matter. For we can say, with propriety, that this Little Treatise is truly Spiritual and Essential to the Happiness of Man. Therefore, to those who wish to be Happy, with every good intention we commend this Work to be their constant companion and study, in which, if they persevere, they shall not fail of their desires in the attainment of the True Philosophers' Stone.

The First Part of Book 2

OF ALCHEMY AND ITS DIVINE ORIGIN

- ❖ The Difficulty of Attaining a Perfection in the Art ~
- ❖ What an Adept is ~ Of the Cabala ~ The Rosicrucians, Adeptists
- ❖ The Possibility of Being an Adept ~
- ❖ The Lapis Philosophorum Exists in Nature and Proven by Sufficient Authority ~
- ❖ They are not all Imposters who are Alchemists or Pretend to it ~
- ❖ The Madness of the Schools Proven ~
- ❖ The Foolishness of Their Wisdom ~
- ❖ The Triumph of Chemical Philosophy or the HERMETIC ART Preferable to any Other

It is not necessary here to enter into a long detail of the merits of Alchemical Authors and Philosophers. Suffice it to say, that Alchemy, the grand touchstone of Natural Wisdom, is of Divine Origin. It was brought down from Heaven by the Angel Uriel.

Zoroaster, the First Philosopher, by Fire, made Pure gold from all the Seven Metals. He brought the Sun 10 times brighter from the Bed of Saturn and fixed it with the Moon, who thereby copulating, begot a numerous offspring of an Immortal Nature, a Pure Living Spiritual Sun, burning in the effulgence of its own Divine Light, a seed of the Sublime and Fiery Nature, a vigorous progenitor.

This Zoroaster was the Father of Alchemy, illumined divinely from above. He knew everything, yet seemed to know nothing.

His Precepts of Art were left in Hieroglyphics, yet in such sort that none but the favorites of Heaven ever reaped benefits thereby.

He was the first who engraved the Pure Cabala in the most Pure Gold and when he died, resigned it to his Father "who lives eternally," yet begot him not.

"That Father gives it to his sons, who follow the precepts of Wisdom with Vigilance, Ingenuity and Industry and with a Pure, Chaste and Free Mind."

Hermes, Trismegistus, Geber, Artephius, Bacon, Helmont, Lully and Basil Valentine, have written most profoundly, yet abstrusely; and all declare not the thing sought for. Some say they were forbid; others that they declared it obviously and intelligibly, yet some few little points they kept to themselves.

However far off the main point they lead us, of this be sure – that something valuable is to be drained, as it were, out of each.

Gerber is good; Artephius is better; but Flammel is best of all; and better still than these is the instructions we give. For with them, a Man (following our directions) shall never want Gold; therefore, to be an Adept is possible, but first "seek the Kingdom of God and all these things shall be added unto you."

This is truth incontrovertible and herein lies a vast secret.

"Seek and you shall find."

But remember, whatsoever you ask, that shall you receive.

The Cabala, in its utmost purity, is contained in the many Precepts given in this book. The Cabala enables us to understand, to bring our understanding to act and, by that means, to attain Knowledge. That Knowledge makes us the Children of God. God makes whom He pleases, Adepts in Wisdom.

To be an Adept, according to God's Will, is no contemptible calling. The Noble and Virtuous Brethren of the Rosy Cross, holds this Truth Sacred, that, "Virtue Flies from no Man."

Therefore, how desirable a thing is Virtue.

She teaches us, first, Wisdom, then Charity, Love, Mercy, Faith and Constancy. All these appertain to Virtue. Therefore, it is physically possible for any well-inclined Man to become an Adept, provided he lays aside his Pride of reasoning, all obstinacy, blindness, hypocrisy, incredulity, superstition, deceit, etcetera.

An Adept, therefore, is one who not only studies to do God's Will upon Earth, in respect of his Moral and Religious Duties, but who studies and ardently Prays to his Benevolent Creator to bestow on him Wisdom and Knowledge from the fullness of His Treasury. And he meditates, day and night, how he may attain the True Aqua Vita; how he may be filled with the Grace of God; which, when he is made so happy, his Spiritual and Internal Eve is open to a glorious prospect of Mortal and Immortal Riches. He wants not food, raiment, joy or anything. He is filled with the Celestial Spiritual Mana. He enjoys the Marrow and Fat Things of the Earth. He treads the Winepress, not of the Wrath, but of the Mercy of God. He lives to the Glory of God and dies saying, "Holy, Holy, Holy Lord of Sabbath! Blessed is your Name, now and forevermore! Amen."

Therefore, to be an Adept, as we have before hinted, is to know yourself, fear God and love your neighbor as yourself; and by this you shall come to the fulfillment of your desires, oh Man. But by no other means under the Scope of Heaven.

When your Soul shall be made drunk by the Divine Ambrosial Nectar, then shall your understanding be more clear than the noontide Sun. Then, by your strong and spiritualized intellectual Eye, you shall see into the Great Treasury of nature and you shall praise God with you whole Heart. Then will you see the Folly of the World and you shall, without Error, accomplish your Desire and (you) shall possess the True Philosopher's Stone, to the profit of your neighbor.

I say, you shall visibly and sensibly, according to your Corporal Faculties, not imaginary, not delusively, but real.

Helmont, an Author of no mean repute, avouches that he has actually seen the Stone which converts Base Metals into Gold, that he has seen it with his Eyes and handled it with his Fingers. (He has) taken from his own relation of the fact. Notwithstanding, Kircher's declamation against the possibility of obtaining it, noting them all who professed Alchemy to be a set of imposters and jugglers, giving no better an exposition of their process of Transmutation than this.

"An Alchemist," says Kircher, "procures or desires a Crucible to be brought, wherein is put Lead or any other Base Metal, which, while in fusion, he (the Alchemist), stirs about with an Iron Rod and then," he says, "he drops in, from between his fingers, a bit of Gold and after stirring up for some time and essay being made, Gold is found."

This is, indeed, a very lame method of exploding Alchemy.

But, however, to leave Kircher as much in the dark as he was, we shall give you Van Helmont's declaration, a Philosopher of much greater note than this Pseudo-Chemist Kircher.

Van Helmont says.

"I have diverse times handled that Stone with my hands and have seen a real Transmutation of saleable Quicksilver with my eyes, which, in proportion, did exceed the Powder which made the Gold in some 1000 degrees.

"It was of the color that is in Saffron, being weighty in its Powder and shining like bruised Glass, when it should be the less exactly beaten. But there was once given to me, the fourth part of one grain. (I call it also a grain the 600th part of an ounce.)

"This Powder I involved in Wax, scraped off a certain letter, lest, in casting it into the Crucible, it should be dispersed, through the Smoke of the Coals, which Pellet of Wax, I afterwards cast into the 3 cornered vessel of a Crucible upon a pound of Quicksilver, hot and newly bought. And presently, the whole Quicksilver, with some little noise, stood from flowing and resided like a lump. But the Heat of that Argent Vive was as much as might forbid melted Lead from re-coagulating. The Fire being straightway after increased under the bellows, the Metal was melted. The which, the vessel of fusion being broken, I found to weight 8 ounces of the most Pure Gold."

(We are to understand in the Alchemist Sense, as differential of the Philosopher's Sense itself, in which the prize of all attainment is a perfect stone of original desire to achieve a magical wit of art; that, the assignment is to produce a Loadstone of Gold from a base metal by the use of wax transformation with other chosen substance and so to prove transubstantiate matter in the environ of astrophysical ether.

Though they had no exact perfect steel of the same quality as our refined, tested polymer and acrylic steels; polymers are achieved for the record by the combination of a reducible element which accounts in the area of a melted wax by texture when melted and also the occurrence of a testable hard metal and the combination of a liquid resolution in the silicone classes, more ethically and otherwise in the corrosive metal classes; they are made together to form a manufactured compound which achieves a kind of natural gum and so become useful; there are many combinations.

But I shall continue with my earlier challenge. And reforming these into a mixture of phosphorous broth, a broth made by introducing halides (helium and neon and bismuth components) into the phosphorous, mixing them until they are literal junk. Herein I consider there must be magic in etherization by its principles. And producing at last this etheric substance to produce the metal tap again by the presence of glass and wax or gypsum. There is no quicksilver because I have used iron and steel which should achieve it; quicksilver is sometimes a precious substance today. And if the iron-steel must be enhanced I will use a cheaper escheat method to add chromium to it.

By Magical Principles alone and the mystery of whatever is this ether, which I assume to be the exhaustive properties of oxides in nitrous-carbonates, as is considered the primary domain of dense space and a substance too long to achieve in explanation here – there should be some kind of gold or at least galena which the present world is in severe need of to restore the balances of stone gravities to the land generally. Someone may let me know someday of this. I consider it more Libor to the Apocalyptic, Cataclysmic, Apocryphal and Pseudepigraphia Ages caught together by means of a desperation.)

"Therefore, a computation being made, a grain of that Powder does convert 19,200 grains of impure and volatile Metal, which is obliterated by the Fire, into True Gold. For that Powder, by uniting the aforesaid Quicksilver unto itself, preserved the same, at one instant, from an Eternal Rust, putrefaction, Death and torture of the Fire, howsoever most violent it was and made it as an Immortal Thing, against any vigor or industry of Art and Fire; and trans-changed it into the Virgin Purity of Gold. At least-wise, one only Fire of Coals is required herein."

By which we see that so learned and profound a Philosopher as Van Helmont could not so easily have been made to believe that there existed a possibility of Transmutation of Base Metals into Pure gold, without he had actually proved the same by experiment. Again, let the standing monuments of Flammel's liberal bounty to the poor, through this mean, to be seen at Paris every day, stand as a testimony to the Truth of the existing possibility of Transmutation.

Likewise, Helmont mentions a Stone that he saw and had in his possession, which cured all disorders, the plague not excepted.

I shall relate the circumstance in his own words, which are as follow.

"There was a certain Irishman, whose name was Butler, being some time great with James, King of England, he being detained in the Prison of the Castle of Vilvord. And taking pity on one, Baillius, a certain Franciscan Monk, a most famous preacher of Gallo-Britain.

(Is there a reason this place is not ever understood for Normandy; or is Gallo-Britain the alter-universe which contains all this wealth of knowledge still occult to ourselves, as in the sense of the wars of the Aquitaine which divided the continent of Europe to lands of prior fables, never to understand the doctrine of it by any modern geographer or fool who reads; and nevertheless to begin again).

And taking pity on one, Baillius, a certain Franciscan Monk, a most famous Preacher of Gallo-Britain (Normandy) who was also imprisoned, having an erysipelas in his arm. On a certain evening, when the Monk did almost despair, he swiftly tinged a certain little Stone in a spoonful of Almond Milk and presently withdrew it thence.

So he says to the Keeper, "Reach this supping to that Monk and how much soever he shall take thereupon, he shall be whole, at least within a short hour's space."

Which thing, even so came to pass, to the great admiration of the Keeper and the Sick Man, not knowing from whence so sudden health shone upon him, seeing that he was ignorant that he had taken anything; for his left arm, being before hugely swollen, fell down as that it could scarcely be discerned from the other. On the morning following, I, being entreated by some great men, came to Vilvord, as a witness of his deeds; therefore, I contracted a friendship with Butler.

Soon afterwards, I saw a poor Old Woman, a Laundress, who, from the age of 16 years, had labored with an intolerable megrim (migraine), cured in my presence. Indeed he, by the way, lightly dipped the same little Stone in a spoonful of oil of olives and presently cleansed the same Stone by licking it with his Tongue and laid it up into his snuff box. But that spoonful of oil he poured into a small bottle of oil, whereof one only drop he commanded to be anointed on the head of the aforesaid Old Woman, who was thereby straightway cured and remained whole; which I attest I was amazed, as if he was become another Midas."

But he, smiling, said (thus).

"My most dear friend, unless you come hitherto, so as to be able, by one only remedy, to cure every disease, you shall remain in your young beginnings, however old you shall become."

I easily assented to this, because I had learned that from the Secrets of Paracelsus and being now more confirmed by sight and hope. But I willingly confess, that that new mode of curing was unaccustomed and unknown to me.

I therefore said, that a young Prince of Our Court, Viscount of Gaunt, Brother to the Prince of Episuoy, of a Very Great House, (Leon or Orleans or also, Orly) was so wholly prostrated by the gout, that he thenceforth lay only on one side, being wretched and deformed with many knots.

He, therefore, taking hold of my right hand said, "Will you that I cure the young man? I will cure him for your sake."

(And I replied.) "But, he is of that obstinacy, that he had rather die than drink one only medicinal potion."

"Be it so," said Butler, "for neither do I require any other thing, than that he do, every morning, touch this little stone (that) you see, with the top of his tongue, for after 3 weeks from thence, let him wash the painful and unpainful knots with his own urine (salts) and you shall soon afterwards see him cured and soundly walking. Go your ways and tell him, with joy, what I have said."

I therefore, being glad, returned to Brussels and told him what Butler had said.

But the Potentate answered, "Go, tell Butler that if he shall restore me as you have said, I will give him as much as he shall require. Demand the price and I will willingly sequester that which is deposited for his security."

And when I declared the thing to Butler, on the day following, he was very wroth and said, "That Prince is mad or witless and miserable and therefore I will never help him. For neither do I stand in need of his money, neither do I yield, nor am I inferior to him."

Now could I ever induce him, afterwards, to perform what before he had promised; wherefore I began to doubt whether the things I had before seen were dreams.

It happened in the meantime, that a friend, overseer and Master of the Glass Furnace at Antwerp, being exceeding fat, most earnestly requested of Butler that he might be freed from his fatness; unto whom Butler offered a small piece of that little Stone, that he might once every morning lick or speedily touch it with the top of his Tongue. And, within 3 weeks, I saw his breast made more straight or narrow, by one span and him to have lived no less whole afterwards.

Wherefore, I began again to believe that the aforesaid gouty Prince might have been cured, according to the manner Butler had promised. In the meantime, I sent to Vilvord, to Butler, for a remedy, in the case of Poison given me by a secret enemy. For I miserably languished; all my joints were pained and my pulse, vehement, being at length become an intermitting one, did accompany the faintings of my mind and extinguishment of my strength.

Butler, being still detained in prison, commanded my Household Servant, whom I had sent, that forthwith he should bring unto him a small bottle of oil of Olives. And his little Stone, aforesaid, being tinged therein, as at other times, he sent that oil to me and told the Servant, that with one only small drop of the oil, I should anoint only one place of the pain or all the places if I would; the which I did and yet felt no help thereby. In the meantime, my enemy, according to his lot, being about to die, bade that pardon should be craved of me for his sin. So I knew that I had taken poison, the which I suspected and therefore, also, I procured with all care to extinguish the slow venom, which, through the grace of God favoring me, I escaped.

Seeing that, afterwards, many other cures were performed upon certain Gentle-Women, I asked Butler why so many Women should be cured, but that I (while that I sharply conflicted with Death itself, being also environed with pains of all my joints and organs) should not feel any ease?

But he asked with what disease I had labored? And when he understood that poison had given a beginning to the disease, he said, that, as the cause had come from within to without, the oil ought to be taken into the body or the stone to be touched with the Tongue, because the grief, being cherished within, it was not local or external. And (he) also observed, that the oil did, by degrees, unclothe itself with the efficacy of healing, because the little Stone, being lightly tinged in it, it had not pithily charged the oil throughout its whole body, but had only ennobled it with a delible (erasable) or obliterable besprinkling of its odor; for truly that Stone did present, in the eyes and tongue, Sea Salt spread abroad or rarified and it is sufficiently known that Salt is not to be very intimately mixed with oil.

This same man, also, cured an abscess, who, for 18 years, had had her right arm swelled with an entire deprivation of motion and the fingers thereof stiff and immovable, only by the touching of her tongue with this admirable Stone.

But very many being present witnesses of these same Wonders, did suspect some Hidden Sorcery or Diabolical Craft, for the Common People have it for an Ancient Custom, that whatsoever honest thing their ignorance has determined not to comprehend, they do, for a privy shift of their ignorance, refer the same to be the juggling of an Evil Spirit. But I could never decline so far, because the remedy was supposed to be Natural. For neither words, ceremonies, nor any other suspected thing, was required. For neither it is lawful, according to Man's Power of Understanding, to refer the Glory of God shown forth in Nature, unto the Devil.

For none of those people had required aid of butler, as from Necromancy anyway suspected. Affirmatively, the things was at first made trail of with smiling and without faith and confidence. Yet this easy method of curing shall long remain suspected by many. For the wit of the Vulgar being inconstant and idle, they do more readily consecrate so great a bounty of restitution unto Diabolical Contrivance, than to Divine Goodness, the framer, lover, savior, refresher of Human Nature and the Father of the Poor.

And these vile prejudices are not only inherent in the Common People, but also in those that are learned, who rashly search into the Beginning of Healing, being not yet instructed or observing the common and blockish rules, because they are always wise as children, who have never gone over their Mother's Threshold, being afraid of every Fable. For they who have not hitherto known the whole circuit of diseases to be included within the Spirit of Life, which makes the assault, or if they hereafter, reading my studies by the way, shall imprint on themselves this moment or concernment of healing; nevertheless, because they have been already before accustomed from the very beginnings of their studies, to the precepts of the humorists, they will easily, at length, depart from me and leap back to the favorite bigotry and ancient opinions of the schools.

But now we will hasten to the manner of preparation necessary to qualify a Man for the attainment of these sublime gifts.

(So then as for our own applications of a simple man-made little Stone, as we want to understand what this stone was that it was man-made in order to be taken into the body or approached by the body; we could guess that it can be coagulated with other necessary substances to the body, known to be curative of some ailment. As he says, olive oil, almond milk, not salt, but considerably salt perhaps to persons with dissuasions of completions of organic substances and so on. I do recommend such things as cured milks, cured salts, cured nut spreads, cured oils, cured refined flours, cured herbs, cured juices and so on, because there are some necessary mineral substances that seem magical to the body like a miraculous medicine and they cannot be digested into the viscera without some means and so this little Stone must have proved itself an aesthetic and traveling antibody and also it must be kept so as not to re-infect the body with its malice after its work. He says nothing of it.)

OF THE PREPARATION OF A MAN TO QUALIFY HIM FOR THE SEARCH OF THIS TREASURE AND OF THE FIRST MATTER (PRIMA MATERIA) OF THE STONE

LESSON ONE

The preparation for this Work is simply this.

Learn to cast away from you all Vile Affections, all levity and inconstancy of Mind.

Let all your dealings be free from deceit and hypocrisy.

Avoid the company of vain young men.

Hate all profligacy and profane speaking.

LESSON TWO

Keep your own and your neighbors' secrets.

Court not the favors of the rich.

Despise not the poor, for he who does will be poorer than the poorest.

LESSON THREE & LESSON FOUR

Give to the need and unfortunate what little you can spare, for he that has but little, whatever he spares to the miserable, God shall amply reward him.

Be merciful to those who offend you, or who have injured you, for what must that man's heart be, who would take heavy vengeance on a slight offense? You shall forgive your brother until 70 times seven.

LESSON FIVE

Be not hasty to condemn the actions of others, unless you should, the next hour, fall into the very same Error.

Despise scandal and tattling. Let your words be few.

LESSON SIX

Study day and night with supplications to the Creator that He would be pleased to grant you Knowledge and understanding.

Pray that the Pure Spirits may have communication and influence in your spirit and soul.

LESSON SEVEN

Be not overcome with drunkenness.

Be assured that half the Evils that befall Mankind originate in drunkenness.

For too great a quantity of strong liquors deprive men of their Reason; then, having lost the use of the Faculty of their Judgment, they immediately become the recipient of all Evil Influences and are justly compared to weather-cocks, that are driven hither and thither by every gust of wind.

So those who drown the Power of Reason, are easily persuaded to the lightest and most frivolous pursuits and, from these, to Vices more gross and reprobate.

The Ministers of Darkness have never so favorable an opportunity of insinuating themselves into the minds and hearts of men, as when they are lost in intoxication.

I pray you to avoid this dreadful Vice.

LESSON EIGHT

Avoid Gluttony and all excess.

Excess if very pernicious and from the Devil.

These are things that constantly tempt Man and by which he falls a prey to his Spiritual Adversary.

For he is rendered incapable of receiving any Good or Divine Gift. Besides, the Divine and Angelic Powers or Essences delight not to be conversant about a Man who is defiled and stinking with debauchery and excess.

LESSON NINE

Covet not much Gold, but learn to be satisfied with enough.

For to desire more than enough is to offend the Deity.

LESSON TEN

Read often these 10 preparatory lessons to fit yourself for the great work and for the receiving of higher things.

For the more pure you are in heart and mind, by so much quicker shall you perceive those High Secrets we teach and which are entirely hid from the discernment of the vicious and depraved, because it never can happen that such a Source of Treasure can be attained merely to satisfy our more gross, earthly and vain desires and inclinations, because here nothing must be thought to be grasped or wrested out of this book, but to the fulfilling of a good end and purpose.

When you shall have so far purified your heart, as we have spoken is indispensably necessary for the receiving of Every Good Thing, you shall then see with other eyes than you do at present.

Your Spiritual Eyes will be opened and you shall read Man as plain as you will our books.

But, for all this, depend not on the strength of your own Wisdom, for even then, when we think our hearts secure, if we do not watch them that they sleep not, the Devil, or his Ministers, immediately take us at this unguarded moment and tempts us into the actual commission of some Sin or other.

Either he excites our appetite for Lust and Concupiscence, or any other Deadly Sin; therefore, using Our Blessed Redeemer's Words, "What I say unto you, I say unto you all ~ WATCH!"

Perhaps, I do not doubt but, there are sine that will say, when they look at our works, this fellow is all rant, all preaching -- he tells us what we knew before as well as himself. To such I say, let them read our book but twice. If they do not gather something that they will acknowledge precious (nay, be convinced that it is precious, to their own satisfaction) I will burn these writings and they shall be no more remembered by me.

To conclude this Part. We say that the First matter (PRIMA MATERIA), Adam brought with him out of Paradise and left it, as an inheritance, to us his successors. Had he remained in his original purity, he would have been permitted to have used it himself; but the Eternal Fiat was passed, that he was to "earn his bread by the sweat of his brow;" therefore, he could not effect what was afterwards performed by some of his offspring.

Hermes Trismegistus, that Ancient Philosopher, wrote touching the attainment of this Stone, which he pronounced to be of all benefit to Man and one of the greatest blessings he could possess. And although his writings contain much of the Excellency of Truth, being wrapped up in such symbolical figures, it renders them exceedingly difficult to be understood, yet, if comprehended, they, no doubt, contain some very great secrets by which Mortal Man may profit.

Now it belongs to our purpose to know what it is from which we must extract the First Matter of this Stone, to go on with our process, because we must have materials to work upon. For all Philosophers agree that, the First Matter being found, we may proceed without much difficulty.

For the First Matter (I shall speak as plainly as possible).

First, the Grand Question in debate is ~ Where is it to be found?

I say it is to be found in ourselves.

We all possess the First Matter, from the Beggar to the King. Every Mother's Son carries it about him. And, could our ingenious chemists but find a process for the extracting, how well would all their labors be repaid.

The Next Question naturally comes to us ~ How are we to draw, or attract the Secret Matter of the Stone out of ourselves?

Not by any Common Means.

Any yet it is to be drawn into very action and that by the most simple means and in a manner that the attaining of the Philosopher's Stone would very soon follow it. I pray you, my Friend, look into yourself and endeavor to find out in what part of your composition is the Prima Materia of the Lapis Philosophorum or out of what part of your Substance can the First Matter of our Stone be drawn out.

You say it must either be in the Hair, Sweat or Excrement. I say in none of these you shall ever be able to find it and yet you shall find it in yourself.

Many Great Philosophers and Chemists, whom I have the pleasure to know, affirm that, admitting of the possibility of Transmutation, it (i.e., the First Matter) must be taken from the purest Gold.

To this I say it must not. Neither has it anything at all to do with extrinsical Gold. They will say then that the pre ENS of Gold may be drawn from Gold itself. True, it may be so. But then I would ask if they could ever produce more Gold than that out of which the Soul or Essence was extracted. If they have, they have indeed found out a secret beyond the powers of our comprehension. Because it is against Reason to suppose that if a Pound of Gold yields a Drachm of the Soul or Essence, that that only will tinge any more than a Pound of Purified Lead – or (☿) because we have tried various experiments and I have, in some of my first essays, turned both lead and mercury into Good Gold, but no more than that out of which the Soul was extracted.

But, however, not to lose our time in vain and ridiculous disputation, know that whatever prodigious things or experiments have been tried with respect to the First Matter, by external subjects, either in the Mineral, Animal or Vegetable Kingdoms, as they are called, I say in us is the Power of all Wonderful Things, which the Supreme Creator has, of his Infinite Mercy, implanted in our Souls, out of her is to be extracted the First Matter, the True Argent Vive, the (☿) of the Philosophers, the True ENS of (☉), viz. a Spiritual Living Gold or waterish Mercury, or FIRST MATTER, which, by being matured, is capable of Transmuting 1000 points of Impure Metal into Good and Perfect Gold, which endure Fire, Test or Cupel.

(Cupel: a shallow, porous container in which gold or silver can be refined or assayed by melting with a blast of hot air, which oxidizes lead or other base metals.)

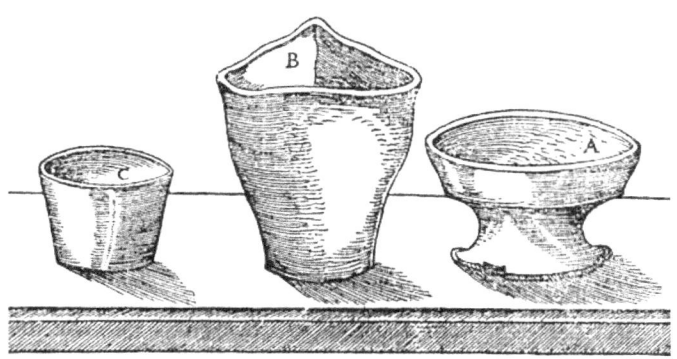

PART TWO of the JEWEL of ALCHEMY

Of the Manner of Extracting THE FIRST MATTER
Of the PHILOSOPHER'S STONE
And the use it is put to in purifying the IMPERFECT METALS
And Transmuting them into GOOD GOLD

LESSON ELEVEN

Take the foregoing instructions as your principal instrument and know that our Soul has the Power, when the Body is Free, as we before said, of any pollution, the Heart Void of Malice and Offense. I say the Soul is then a Free Agent and has the Power, spiritually and magically, to act upon any matter whatsoever; therefore I said the First Matter is the Soul and the extracting of it, is to bring the dormant power of the pure, living, breathing spirit and Eternal Soul into act.

Note well that every agent has its power of acting upon its patient. Every Essence that is distilled forth is received into a recipient, but that recipient must first be made clean. Even so must the Soul and Heart of Man; the Vile Affections must be thrown away and trampled underfoot. Then shall you be able to proceed in your work, which (is done) in the following manner.

LESSON TWELVE

The expense that you must have will be but a trifle; all the Instruments necessary are but 3, viz. a Crucible, an Egg Philosophical and a retort with its receiver.

Put your Fine Gold, in weight of about 5 (dwts.) pennyweights, file it up, put it into your Philosophic Egg, pour upon it the twice of its weight of the best Hungarian (☿), close up the Egg with a Hermetic Seal, put it for 3 months in Horse-dung, take it out at the end of that time and see what kind of form your gold and (☿) has assumed.

Take it out, pour on it half its weight of Good Spirit of Sal-Ammon, set them in a pot full of Sand over the Fire in the retort, let them distill into a pure essence, add to one pint of this (☿) two pints of your Water of Life, or Prima Materia, put them into your Philosophical Egg and set them into Horse-dung for another 3 months.

(When you have set them into the Horse-dung for another 3 months, you have begun on Lesson Thirteen.)

LESSON THIRTEEN

(Continuing on the Exercise of Lesson Twelve in direction of current timing.) Then take them out and see what you have; a Pure Ethereal Essence, which is the Living Gold.

Pour this Pure Spiritual Liquor upon a Drachm of Molten Fine Gold and you will find that which will satisfy your hunger and thirsting after this Secret.

Take it to a Jeweler's or Goldsmith's.

Let him try it in your presence and you will have reason to Bless Bod for His Mercy to you.

Do your duty as He has commanded you ad use all the benefit you shall receive, in actions worthy of your Nature.

LESSON FOURTEEN

When your Spiritual Eve is opened and you shall begin to see to what end you were created, you shall want no necessary thing, either for your comfort or support.

Only keep in the rules we have prescribed in the Beginning of this Little Treatise. Fear God and Love your Neighbor as yourself! Be not hasty to reveal any Secrets you may learn, for the Good Spirits, both Day and Night, will be your instructors and will continually reveal to you many Secrets.

Think not that you can either profit or benefit so much by the Instruction of those who profess great advantages in Classical Education and High Schooling. Be assured they are, in Spiritual Knowledge, much in the Dark.

For he who desires not Spiritual Knowledge cannot attain it by, any means, but by, first, coming to God; secondly, by purifying his own Heart; thirdly, by submitting himself to the Will of the Holy Spirit, to guide and direct him in all Truth, to the attaining of all Knowledge, both Human and Divine; and by arrogating Nothing to our own Power or Strength, but by referring to all the Mercy and Goodness of God.

Amen. (All say Amen for the gift of the objective of receiving a Divine Promise from proper Instruction and the request of Enlightenment.)

PART 3

THE MAGUS, Or, CELESTIAL INTELLIGENCER

Containing the Constellatory Practice

Or, TALISMANIC MAGIC

Showing the true properties of the Elements, Meteors, Stars, Planets, etcetera, etcetera; likewise the Nature of Intelligences, Spirits, Demons and Devil, the construction and composition of all sorts of Magic Seals, Images, Rings, Glasses, Pictures, etcetera, etcetera; the power and composition of Numbers, Mathematical Figures and Characters of Spirits, both Good and Evil

The Whole of the Above Illustrated by a Great Variety of Beautiful Figures, types, Letters, Seals, Images, Magic Characters, etcetera, forming a Complete System of Delightful Knowledge and Abstruse Science, such as is warranted never before to have been published in the English language.

By Francis Barret, Student of Chemistry, Occult Philosophy and the Cabala, 1801

This is the Second Part of THE JEWEL OF ALCHEMY ~
Being also the Second Part of the Greater First BOOK of THE MAGUS

Chapter 1
Of the Four Elements and their Natural Qualities
(from where we derive the Perfection of the Secret Cabala)

It is necessary that we should know and understand ~

the NATURE and QUALITY of the FOUR ELEMENTS ~
in order to our BEING PERFECT in the Principles
and Groundwork of our Studies in the ~

TALISMANIC or MAGICAL ART.

Therefore, there are FOUR ELEMENTS ~
the ORIGINAL GROUNDS of ALL CORPOREAL THINGS ~ viz.

> FIRE
> EARTH
> WATER
> AIR

Of which ELEMENTS, all Inferior Bodies are compounded, not by way of being heaped up together, but by Transmutation and Union.

And when they are destroyed, they are resolved into Elements.

But there are none of the Sensible Elements that are Pure.

But they are, more or less, mixed and apt to be change, the one into the other.

Even as EARTH, being moistened and dissolved, becomes WATER, but the same, being made thick and hard, becomes EARTH again.

And being evaporated through heat it passes into AIR and that being kindled into FIRE and this being extinguished, into AIR again, but being cooled after burning, becomes EARTH again or else Stone, or Sulphur; and this is clearly demonstrated by Lightning.

Now every one of these ELEMENTS have two specific properties.

The former whereof, (or, being in the former of itself whereof) it retains (itself) as a property to itself and in the other, (in the next form, it retains a mean of itself) as a mean, (and) it agrees with that which comes directly after it.

(The Four Base Elements of the Creative Physical and Natural Universe is mutable to the end that it takes a physical (if not also astrophysical) effect upon itself in changeable forms (as shown). And furthermore, each of the Four Principles Elements out of which all other Elements are derived, are common in quality of usage the same, that if it begins in the principle combination with another partner, it remains the principle Element and becoming into participle with the next Element, it becomes a vehicle of change to agree again with the next Element that results after it. In this way, there never results another purity of elements created after those Four Base Substances.
    And this is because, as shown below of their self-inherent substantial nature and qualities that are contrary and active to and with one another.)

(For) FIRE is hot and dry; EARTH is cold and dry; WATER is cold and moist; and AIR, hot and moist.

And so in this matter, the ELEMENTS, according to two contrary qualities, are opposite one to the other, as FIRE to WATER and EARTH to AIR.

Likewise, the ELEMENTS are contrary one to the other in another account.

Two are heavy, as EARTH and WATER; and the others are light, as FIRE and AIR.

Therefore, the Stoics called the former, passives, but the latter, actives. (EARTH and WATER are passive and FIRE and AIR are active).

And PLATO distinguishes them after another manner and assigns to each of them 3 qualities, viz. to the FIRE ~ brightness, thinness and motion; to the EARTH ~ darkness, thickness and quietness; and, according to these qualities, the ELEMENTS of FIRE and EARTH are contrary (as we can easily understand by their opposite assignments).

Now the other ELEMENTS borrow their qualities from these, so that the AIR receives 2 qualities from the FIRE – thinness and motion; and the EARTH (receives) one – darkness.

In like manner, WATER receives 2 qualities of the EARTH – darkness and thickness; and the FIRE (receives) one – motion.

But FIRE is twice as thin as AIR, three times more moveable and four times brighter.

The AIR is twice more bright, three times more thin and four times more moveable.

Therefore, as FIRE is to AIR, so AIR is to WATER and WATER to the EARTH.

And again, as the EARTH is to the WATER, so WATER is to AIR and AIR to FIRE.

And this is the root and foundation of all bodies, natures and wonderful works. And he who can know and thoroughly understand these qualities of the ELEMENTS and their mixtures, shall bring to pass wonderful and astonishing things in MAGIC.

Now each of these ELEMENTS have a three-fold consideration, so that the number of FOUR makes up the number of TWELVE.

And, by passing by the number of SEVEN into TEN, there may be a progress to the SUPREME UNITY upon which ALL VIRTUE and wonderful things do depend.

OF THE FIRST ORDER ARE THE PURE ELEMENTS, which are neither compounded, changed or mixed, but are INCORRUPTIBLE.

And not OF WHICH, but THROUGH WHICH, the VIRTUES OF ALL NATURAL THINGS are brought forth to act.

No Man is able fully to declare their Virtues, because they can do all things upon all things.

He who remains ignorant of these, shall never be able to bring to pass any wonderful matter.

OF THE SECOND ORDER are ELEMENTS that are COMPOUNDED, changeable and IMPURE.

Yet such as many, by art, may be reduced to their PURE simplicity, whose VIRTUE, when they are thus reduced, do, above all things, perfect all occult and common operations of Nature and these are the FOUNDATION OF THE WHOLE OF NATURAL MAGIC.

OF THE THIRD ORDER, are those ELEMENTS which originally and of themselves are not ELEMENTS, but are twice COMPOUNDED, various and changeable into another.

These are the INFALLIBLE MEDIUM and are called the MIDDLE NATURE, or SOUL of the MIDDLE NATURE.

Very few there are that understand the Deep Mysteries thereof.

(In them is, ~)
by means of CERTAIN NUMBERS, DEGREES and ORDERS, the PERFECTIONS of every effect in what thing soever; whether NATURAL, CELESTIAL or SUPER-CELESTIAL.

They are FULL OF WONDERS and MYSTERIES and are operative as in MAGIC NATURAL, so, DIVINE.

For from these, through them, proceeds the BINDING, LOOSING and TRANSMUTATION of ALL THINGS; the KNOWLEDGE and FORE-TELLING of THINGS TO COME.

(Also, ~)
the EXPELLING OF EVIL and the GAINING OF GOOD SPIRITS.

Let no one therefore, without these THREE SORTS OF ELEMENTS and the TRUE KNOWLEDGE thereof, be confident that he can work anything in the OCCULT SCIENCE OF MAGIC and NATURE.

But whosoever shall know how to reduce those of one ORDER into (those of) another, IMPURE to PURE, COMPOUNDED into SIMPLE and shall understand distinctly the Nature, Virtue and Power of them, in Number, Degrees and Order, without dividing the Substance, shall attain to the KNOWLEDGE and PERFECT OPERATION of ALL NATURAL THINGS and CELESTIAL SECRETS likewise.

And this is the PERFECTION OF THE CABALA, which teaches all these before-mentioned; and, by a PERFECT KNOWLEDGE thereof, we perform many rare and wonderful experiments.

Chapter 2
Of the Properties and Wonderful Nature of FIRE and EARTH

There are two things, says Hermes, viz. FIRE and EARTH, which are sufficient for the Operation of All Wonderful Things.

FIRE is active and EARTH is passive.

FIRE, in all things and through all things, comes and goes away brightly. It is in all things bright and at the same time, Occult and Unknown. When it is by itself (no other Matter coming to it, in which it should manifest its proper action) it is proper boundless and invisible. Of itself, sufficient for every action that is proper to it.

(Of itself it remains ONE (thing)) and penetrates through all things; also, spread abroad into the Heavens, it shines alone. But in the Infernal Places, it is straightened and dark and tormenting. And in the Midway, it partakes of both (qualities of its own essence).

It is (placed or found) in Stones and drawn out by the stroke of the Steel. It is (placed or found) in the Earth and causes it, after digging up, to smoke. It is (placed or found) in Water and it heats it and springs and wells it. It is (placed or found) in the Depths of the Sea and causes it, being tossed with the Winds, to be hot. It is (placed or found) in the Air and makes it (as we often see) to burn.

And all Animals and all Living Thing whatsoever, as also Vegetables, are preserved by heat. And everything that lives, lives by reason of the enclosed heat.

The Properties of Fire are a Parching Heat, consuming all things and Darkness, making all things barren.

The Celestial and Bright Fire drives away Spirits of Darkness. Also, this, our Fire, made with wood, drives away the same, in as much as it has an analogy with and is the vehicle of, that Superior Light.

As also of him who says, "I am the Light of the World," which is the True Fire; the Father of Lights, from whom every good thing that is given, comes, sending forth the Light of His Fire and communicating it first to the Sun and the rest of the Celestial Bodies and by these, as by mediating instruments, conveying that Light into our Fire.

As therefore, the Spirits of Darkness are stronger in the Dark, so Good Spirits, which are Angels of Light, are augmented not only by that Light (which is Divine, of the Sun and Celestial), but also by the Light of the our Common Fire.

Hence it was that the first and most wise institutors of religions and ceremonies, ordained that prayers, singing and all manner of Divine Worship whatsoever, should not be performed without lit candles or torches.

Hence, also, was the significant saying of Pythagoras.

"Do not speak of God without a light!"

And they commanded that, for the driving away of Wicked Spirits, lights and fires should be kindled by the Carcasses of the Dead and that they should not be removed until the expiations were, after a holy manner, performed and then buried.

And the Great Jehovah himself, in the Old Law, commanded that all his sacrifices should be offered with Fire and that fire should always be burning upon the Altar, which custom the Priests of the Altar did always observe and keep amongst the Romans.

Now the Basis and Foundations of all the Elements is the Earth.

For that is the object, subject and receptacle of all Celestial Rays and influences; in it are contained the seeds and Seminal Virtues of all things and therefore, it is said to be Animal, Vegetable and Mineral.

It, being made fruitful by the other Elements and the Heavens, brings forth all things of itself.

It receives the Abundance of All Things and is, as it were, the First Fountain from whence all things spring. It is the center, foundation and mother of all things.

Take as much of it as you please, separated, washed, depurated and subtilized and if you let it lie in the open air a little while, it will, being full and abounding with Heavenly Virtues, of itself brings forth Plants, Worms, and other Living Things, and also Stones and Bright Sparks of Metals.

In it are Great Secrets. If at any time it shall be purified, by the help of Fire and reduced into its simple Nature by a convenient washing, it is the First Matter of Our Creation and the truest Medicine that can restore and preserve us.

Chapter 3
Of the WATER and AIR

The other TWO ELEMENTS, viz. WATER and AIR, as not less efficacious than the former, neither is Nature wanting to work wonderful things in them.

There is so great a necessity of Water, that without it, nothing can live; no Herb nor Plant whatsoever without the moistening of Water, can bring forth. In it is the Seminary Virtue of All Things, especially of Animals, whose seed is manifestly waterish.

The Seeds, also, of Trees and Plants, although they are Earthy, must, notwithstanding, of necessity be rotted in Water before they can be fruitful. Whether they be imbibed with the Moisture of the Earth, or with Dew or Rain, or any other Water that is on purpose put to them.

For Moses writes, that only Earth and Water can bring forth a Living Soul, but he ascribes a two-fold production of things to Water, viz. of things swimming in the Water and of things flying in the Air above the Earth. And that those productions that are made in and upon the Earth as partly attributed to the very Water, the same Scriptures testifies, where it say, that the Plants and the Herbs did not grow, because God had not caused it to rain upon the Earth.

Such is the efficacy of this Element of Water, that Spiritual Regeneration cannot be done without it, as Christ Himself testified to Nicodemus. Very great, also, is the Virtue of it in the religious worship of God, I expiations and purifications. Indeed, the necessity of it is no less than that of Fire.

Infinite are the benefits and diverse are the uses, thereof as being that, by Virtue of which all things subsist, are generated, nourished and increased. Hence it was that Thales of Miletus and Hesiod, concluded that Water was the Beginning of All Things and said it was the First of all the Elements and the most potent. And that, because it has the mastery over all the rest.

For, as Pliny says.

"Waters swallow up the Earth, extinguish flames, ascend on high; and, by the stretching of the clouds, challenge of the Heavens for their own. The same, falling down, becomes the cause of all things that grow in the Earth."

Very many are the wonders that are done by Waters, according to the writings of Pliny, Solinus and many other historians.

Josephus also makes relation of the wonderful nature of a certain river between Arcea and Raphanea, cites Syria, which runs with a full channel all the Sabbath Day and then all of a sudden, stops, as if the springs were stopped and all the Six Days you may pass over it, dry. But again, on the 7th day, no man knowing the reason of it, the waters return again, in abundance as before! Wherefore, the inhabitants thereabout called it the Sabbath Day River because of the 7th Day, which was Holy to the Jews.

The Gospel also, testifies of a Sheep Pool, into which whosoever stepped first after the Water was troubled by the Angel, was made whole of whatsoever disease he had. The same Virtue and efficacy, we read, was in a spring of the Ionian Nymphs, which was in the Territories belonging to the Town of Elis, at a Village called Heradea, near the River Citheron, which whosoever stepped into, being diseased, came forth whole and cured all of his diseases.

Pausanias also reports, that in Lyceus, a Mountain of Arcadia, there was a spring called Agria, to which, as often as the dryness of the region threatened the destruction of fruits, Jupiter, Priest of Lyceus, went; and, after the Offering of Sacrifices, devoutly praying to the Waters of the Spring, holding a Bough of an Oak in his hand, put it down to the bottom of the Hallowed Spring. Then, the Waters being troubled, a Vapor ascending from thence into the Air, was blown into the Clouds, which being joined together, the whole Heaven was overspread. Which, being a little after dissolved into rain, watered all the country most wholesomely.

Moreover, Ruffus, a Physician of Ephesus, besides many other Authors, wrote strange things concerning the wonders of Waters, which, for aught I know, are found in no other Author.

It remains, that I speak of the Air.

This is a Vital Spirit passing through all Being, giving Life and Substance to all things, moving and filling all things. Hence it is that the Hebrew doctors reckon it not amongst the Elements, but count it as a medium, or glue, joining things together and as the resounding Spirit of the World's Instrument. It immediately receives into itself the influence of all Celestial Bodies and then communicates them to the other Elements, as also to all mixed bodies.

Also, it receives into itself, as if it were a Divine Looking-Glass, the Species of All Thing, as well Natural as Artificial. As also of all manner of speeches and retains them. And carrying them with it and entering into the Bodies of Men and other Animals, through their pores, makes an impression upon them, as well when they are asleep as when they are awake and affords matter for diverse strange dreams and divinations.

Hence, they say, it is that a Man, passing by a place where a Man was slain, or the carcass newly hid, is moved with fear and dread, because the Air in that place, being full of the dreadful species of Manslaughter, does, being breathed in, move and trouble the Spirit of the Man with the like Species; whence it is that he becomes afraid.

For everything that makes a sudden impression, astonishes nature, whence it is that many Philosophers were of opinion, that Air is the cause of Dreams and of many other Impressions of the Mind, through the prolonging of Images, or Similitudes, or Species (which proceed from things and speeches, multiplied in the very Air), until they come to the Senses and then to the Fantasy and Soul of him that receives them, which, being freed from cares and no way hindered, expecting to meet such Kind of Species, is formed by them. For the Species of things, although of their own proper Nature, they are carried to the Senses of Men and other Animals in general, may, notwithstanding, get some impression from the Heavens while they are in the Air. By reason of which, together with the aptness and disposition of him that receives them, they may be carried to the Sense of one, rather than of another.

And hence, it is possible, naturally, and from all Manner of Superstition (no other Spirit coming between), that a Man should be able, in a very small time, to signify his mind unto another Man, abiding at a very long and unknown distance from him; although he cannot precisely give an estimate of the time when it is, yet, of necessity, it must be within 24 hours.

And I, myself, know how to do it and have often done it. The same also, in time past, did the Abbot Tritemius, both know and do.

Also, when certain appearances (not only Spiritual, but also Natural) do flow forth from things, that is to say, by a certain kind of flowings forth of bodies from bodies and do gather strength in the Air, they show themselves to us as well through light as motion – as well to the sight as to other sense – and sometimes work wonderful things upon us, as Platonius proves and teaches.

And we see how, by the South Wind, the Air is condensed into thin Clouds, in which, as in a Looking Glass, are reflected representations at a great distance, of Castles, Mountains, Horses, Men and other things, which when the Clouds are gone, presently vanish.

And Aristotle, in his Meteors, shows that a Rainbow is conceived in a Cloud of Air, as in a Looking Glass.

And Albertus says, that the effigies of bodies may, by the Strength of Nature, in a moist Air, be easily represented. In the same manner as the representations of things are in things.

And Aristotle tells of a Man, to whom it happened, by reason of the weakness of his sight, that the Air that was near to him, became, as it were, a Looking Glass to him and the optic beam did reflect back upon himself and could not penetrate the Air, so that, whither soever he went, he thought he saw his own Image, with his face towards him, go before him.

In like manner, by artificialness of some certain Looking Glasses, may be produced at a distance, in the Air, besides the Looking Glasses, what Images we please. Which, when ignorant men see, they think they see the appearances of Spirits or Souls, when, indeed, they are nothing else but semblances akin to themselves and without Life.

And it is well-known, if in a Dark Place, where there is no Light, but by the coming in of a Beam of the Sun somewhere through a little hole, a white paper or plain Looking Glass be set up against the Light, that there may be seen upon them whatsoever things are done without, being shined upon by the Sun.

And there is another slight or trick yet more wonderful. If anyone shall take Images, artificially painted or written letters, and, in a clear night, set them against the Beams of the Full Moon, those resemblances being multiplied in the Air and caught upward and reflected back together with the Beams of the Moon, another man, that is privy to the thing, at a long distance, sees, reads and knows them in the very compass and circle of the Moon.

Which Art of Declaring Secrets is, indeed, very profitable for towns and cities that are besieged, being a thing which Pythagoras long since did and which is not unknown to some in these days. I will not except myself. And all these things and many more and much greater than these, are grounded in the very Nature of the Air and have their reasons and causes declared in Mathematics and Optics. And as these resemblances are reflected back to the sight, so also are they, sometimes, to the hearing, as if manifest in echo.

But there are many more Secret Arts than these and such whereby anyone may, at a remarkable distance, hear and understand distinctly, what another speaks or whispers.

Chapter 4
Of Compound or Mixed Bodies ~
In What Manner they Relate to the ELEMENTS
And how the ELEMENTS relate to the SOULS, SENSES and DISPOSITIONS OF MEN

The next in order, after the Four Simple Elements, are the Four Kinds of Perfect Bodies compounded of them, viz. metals, stones, plants and animals and although in the generation of each of these, all the Elements combine together in the composition, yet every one of them follows and resembles one of the Elements which is most predominant. For all Stones, being Earthy, are naturally heavy and are so hardened with dryness that they cannot be melted. But metals are watery and may be melted, which Naturalists and Chemists find to be true, viz. that they are composed or generated of a Viscous Water or watery Argent Vive.

Plants have such an affinity with the Air, that unless they are out in it and receive its benefit, they neither flourish nor increase.

So also Animals, as the Poet finely expresses it;

Have, in their Natures, a most Fiery Force,
And also spring from a Celestial Source.

And Fire is so Natural to them, that, being extinguished, they soon die.

Now, amongst Stones, those that are dark and heavy, are called Earthy; those which are transparent of the watery Element, as Crystal, Beryl and Pearls; those which swim upon the water and are spongious, as the Pumice Stone, Sponge and Sophus, are called Airy; and those are attributed to the Element of Fire, out of which Fire is extracted or which are resolved into Fire, as Thunder Stones, Fire Stones, Asbestos. Also, amongst Metals, Lead and Silver are Earthy. Quicksilver is Watery. Copper and Tin, are Airy; and Gold and Iron are Fiery. In Plants, also, the Roots resemble Earth; the Leaves resemble Water, the Flowers resemble the Air and the Seed, resembles the Fire, by reason of their multiplying Spirit.

Besides, some are hot, some are cold, some are moist and others are dry, borrowing their names from the qualities of the Elements. Amongst Animals, also, some are, in comparison of others, Earthy, because they live in the very bowels of the Earth, as Worms, Moles and many other Reptiles; others, Watery, as Fish and others, which always abide in the Air, live Airily. And others still live Fiery, as Salamanders, Crickets and such as are of a Fiery Heat, as Pigeons, Ostriches, Eagles, Lions, Panthers and etcetera, etcetera.

Now, in Animals, the Bones resemble Earth. Vital Spirit resembles the Fire and Flesh, resembles the Air. Humors resemble the Water and these Humors also resemble the Elements. As in the Yellow Cholera (as a Humor) resembles the Fire and the Blood resembles the Air; Phlegm resembles the Water and Black Cholera (as a Humor) or Melancholy resembles the Earth. And lastly, in the Soul itself, the Understanding resembles the Fire and Reason resembles the Air and Imagination resemble the Water.

The Senses resemble the Earth. And these Senses again are divided amongst themselves, according to the Elements. For the Sight is Fiery, because it cannot perceive without the help of Fire and Light. The Hearing is Airy, for a Sound is made by the striking of the Air. The Smell and Taste resemble Water, without the Moisture of which, there is neither smell nor Taste. And lastly, the feeling is wholly Earthly, because it takes gross bodies for its object.

The Actions, also and Operations of Man, are governed by the Elements, for the Earth signifies a slow and firm motion. The Water signifies fearfulness, sluggishness and remiss to working. Air signifies cheerfulness and an amiable disposition. But Fire signifies a fierce, working, quick, susceptible disposition.

The Elements are, therefore, the First and Original Matter of All Things and All Things are of and according to them and they in and through All Things, diffuse their Virtues.

Chapter 5
That the Elements are in the Heavens, in the Stars, in Devils, Angels, Intelligences and lastly, in God Himself.

In the Original and Exemplary World, All Things are All in All. So also in the Corporeal World. And the Elements are not only in these inferior things; but are in the Heavens, in Stars, in Devils, in Angels and likewise in God Himself, the Maker and Original Example of All Things.

Now it must be understood that in these inferior bodies the Elements are gross and corruptible, but in the Heavens, they are, with their Natures and Virtues, after a Celestial and more Excellent Manner than in Sublunary Things; for the firmness of the Celestial Earth is there without the Grossness of Water and the agility of Air without exceeding its bounds. The Heat of Fire without burning, only shining, giving Light and Life to All Things by its Celestial Heat.

Now amongst the Stars, or Planets, some are Fiery, as Mars and the Sun – Airy, as Jupiter and Venus – Watery, as Saturn and Mercury – and Earthy, such as inhabit the Eighth Orb and the Moon (which by many is accounted Watery), seeing that, as if it were Earth, it attracts to itself the Celestial Waters, with which being imbibed it does, on account of its proximity to us, pour forth and communicate to Our Globe. There are, likewise, amongst the Signs, some Fiery, some Airy, some Watery and some Earthy.

The Elements rule them, also, in the Heavens, distributing to them these Four Threefold considerations of every Element, according to their Triplicities, viz. the Beginning, Middle and End. Likewise, Devils are distinguished according to the Elements. For some are called Earthy Devils, others are called Fiery Devils; some are called Airy Devils and still others are called Watery Devils.

Hence, also, those Four Infernal Rivers; Fiery Phlegethon, Airy Cocytus, Watery Styx, Earthy Acheron. Also, in the Gospel, we read of comparisons of the Elements ~ as Hell Fire and Eternal Fire, into which the Cursed shall be commanded to go; and in Revelations, of a Lake of Fire.

And Isaiah, speaking of the damned, says that the Lord will smite them with corrupt Air. And in Job, they shall skip from the Waters of the Snow to the Extremity of Heat. And in the same, we read, that the Earth is Dark and covered with the Darkness of Death and miserable Darkness.

And these Elements are placed in the Angels of Heaven and the Blessed Intelligences; there is in them a stability of their Essence, which is an Earthy Virtue, in which is the Steadfast Seat of God.

By the Psalmist they are called Water, where he says, "Who rules the Waters that are Higher than the Heavens."

Also, in them, their subtle Breath, is Air and their love is shining Fire; hence, they are called in Scripture, the Wings of the Wind. And, in another place, the Psalmist speaks of them thus, "Who makes Angels your Spirits and your Ministers of Flaming Fire!"

Also, according to the different Orders of Spirits or Angels; some are Fiery, as Seraphim, Authorities and Powers; some are Earthy, as Cherubim; some are Watery as Thrones and Archangels; and some are Airy as Dominions and Principalities.

And do we not read of the Original Maker of All Things, that the Earth shall be opened and bring forth a Savior? Likewise, it is spoken of the same, that he shall be a Fountain of Living Water, cleansing and regenerating and the same Spirit breathing the Breath of Life and the same, according to Moses' and Paul's Testimony – a Consuming Fire.

That the Elements are, therefore, to be found everywhere and in All Things, after their (own) Manner, no Man will dare to deny. First, in these inferior bodies, feculent and gross; and in Celestials, more pure and clear; but in Super-Celestials, Living and in all respects Blessed. Elements, therefore, in the Exemplary World, are ideas of things to be produced. In Intelligences, they are distributed powers and in the Heavens, they are Virtues and in inferior bodies, are gross forms.

Chapter 6

That the Wisdom of God works by the Medium of Second Causes. That is, that the Wisdom of God works by the Intelligences, by the Heavens, Elements and Celestial Bodies is proven beyond dispute in this Chapter.

- It is to be noted, that God, in the first place, is the End and Beginning of all Virtue.

- He gives the Seal of the Ideas to His Servants, the Intelligences, who, as faithful officers, sign All Things entrusted to them with an Ideal Virtue.

- And the Heavens and the Stars, as Instruments, disposing the Matter, in the meanwhile, for the receiving of those forms which reside in Divine Majesty and to be conveyed by Stars.

- And the Giver of Forms distributes them by the Ministry of His Intelligences, which he has ordained as Rulers and Comptrollers over His Works.

- To whom, such a Power is entrusted, in Things committed to Them, that so All Virtue in Stones, Herbs, Metals and All Other Things, maybe come from the Intelligences, the Governors.

- Therefore, the Form and Virtue of Things comes First from the Ideas – then from the Ruling and Governing Intelligences – then from the Aspects of Heavens disposing – and lastly, from the Tempers of the Elements disposed, answering the Influences of the Heavens, by which the Elements themselves are ordered or disposed.

- These Kinds of Operations, therefore, are performed in these Inferior Things by Express Forms and in the Heavens, by disposing Virtues, in Intelligences, by mediating Rules, in the Original Cause, by Ideas and Exemplary Forms. All which must of necessity agree in the Execution of the Effect and Virtue of Everything.

- There is, therefore, a Wonderful Virtue and Operation in Every Herb and Stone, but Greater in a Star; beyond which, even from the Governing Intelligences, Everything, receives and obtains many Things for Itself, especially from the Supreme Cause, with whom All Things mutually and exactly correspond, agreeing in a Harmonious Consent.

- Therefore, there is no other Cause of the Necessity of Effects, than the Connection of All Things with the First Cause and their Correspondence with those Divine Patterns and Eternal Ideas, whence Everything has its determinate and particular place in the Exemplary World, from whence it lives and receives its Original Being.

And Every Virtue of Herbs, Stones, Metals, Animals, Words, Speeches and All Things that are of God, are placed There (in the Exemplary World).

Now the First Cause (which is God), although he does, by Intelligences and the Heavens, work upon these Inferior Things, does sometimes (these Mediums being laid aside, or their officiating being suspended) work those things immediately by Himself – which works are then called Miracles.

But whereas, Secondary Causes do, by the Command and Appointment of the First Cause, necessarily act and are necessarily act and are necessitated to produce their effects; if God shall, notwithstanding, according to His Pleasure, so discharge and suspend them that they shall wholly desist from the necessity of that command, then they are called the greatest Miracles of God. For instance; the Fire of the Chaldean Furnace did not burn the children. The Sun stood still at the Command of Joshua and became retrograde one whole day. Also, at the Prayer of Hezekiah, it went back 10 degrees and when the Savior, Christ, was crucified, it became darkened, though at Full Moon.

And the reason of these Operations can by no rational discourse, no Magic or Science, Occult or Profound soever, be found out or understood, but are to be Learned by Divine Oracles only.

Chapter 7
Of the Spirit of the World

Now seeing that the Soul is the Essential Form, Intelligible and Incorruptible and is the First Mover of the Body and is moved of itself – but that Body, or Matter, is of itself unable and unfit for Motion and does very much degenerate from the Soul, it appears that there is need of a more Excellent Medium.

Now, such a Medium is conceived to be the Spirit of the World, or that which some call a QUINTESSENCE. (This) because it is not from the FOUR ELEMENTS, but a CERTAIN FIRST THING, having its BEING above and besides them. There is, therefore, such a Kind of Medium required to be, by which Celestial Souls may be joined to Gross Bodies and bestow upon them Wonderful Gifts.

The Spirit is, in the same manner, in the Body of the World, as our Spirit is in our Bodies. For as the Powers of our Soul are communicated to the Members of the Body, by the Medium of the spirit, so also the Virtue of the Soul of the World is diffused, throughout All Things, by the Medium of the Universal Spirit.

For there is Nothing to be found in the Whole World that has not a Spark of the Virtue thereof. Now this Spirit is received into Things, more or less, by the Rays of the Stars, so far as Things are disposed, or made fit recipients of it.

By this Spirit, therefore, Every Occult Property is conveyed into Herbs, Stones, Metals and Animals, through the Sun, Moon, Planets and through Stars higher than the Planets.

Now this Spirit may be more advantageous to us if we knew how to separate it from the Elements; or, at least, to use those things chiefly which are most abounding with this Spirit. For those Things in which the Spirit is less drowned in a Body and less checked by Matter, do much more powerfully and perfectly act and also more readily generate their like.

For in it are All GENERATIVE and SEMINAL VIRTUES.

For which cause, the Alchemist endeavors to separate this Spirit from Gold and Silver, which, being rightly separated and extracted, if it shall be afterwards projected upon any Metal, turns it into Gold or Silver; which is no way impossible or improbable, when we consider that by art that may be done in a short time, what Nature, in the bowels of the Earth (as in a Matrix) perfects in a very long space of time.

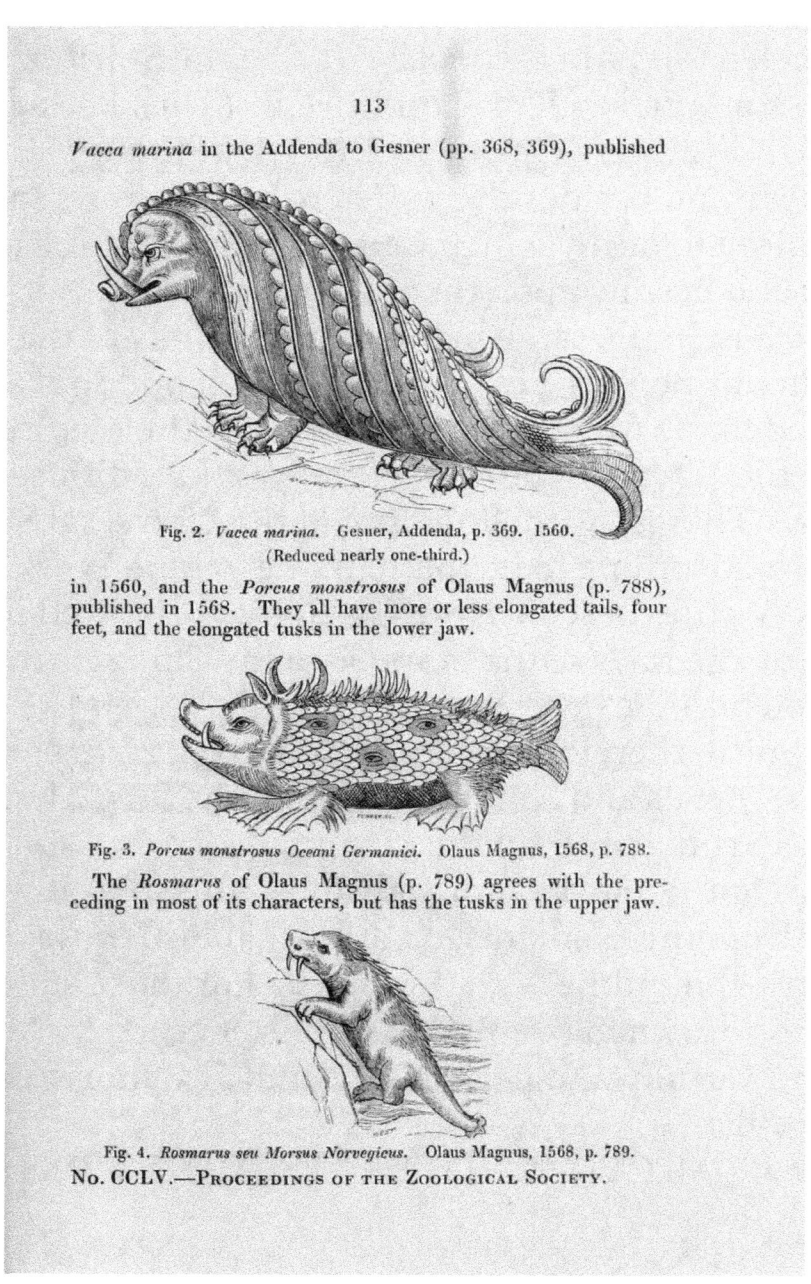

Chapter 8
Of the Seals and Characters
Impressed by Celestials upon Natural Things

All Stars have their peculiar Natures, Properties and Conditions.

The Seals and Characters whereof, they produce through their Rays even in these Inferior Things, viz. in Elements, Stones, in Plants, in Animals and their Members; whence everything receives from a harmonious disposition and from its Star Shining upon it, some particular Seal or Character stamped upon it, which is the significations of that Star or Harmony, containing in it a peculiar Virtue, different from other Virtues of the same Matter, both generically, specifically and numerically.

Everything, therefore, has its character impressed upon it by its Star for some peculiar effect, especially by that Star which does principally govern it. And these characters contain in them the particular Natures, Virtues and Roots of their Stars and produce the like operations upon other things on which they are reflected. And stir up and help the influences of their Stars, whether they are Planets or Fixed Stars and Figures or Celestial Constellations, viz. as often as they shall be made in a fit matter and in their due and accustomed times; which the Ancient Wisemen (considering such as labored much in finding out Occult Properties of Things) did set down, in writing, the Images of the Stars, their Figures, Seals, Marks, Characters, such as Nature herself did describe by the Rays of the Stars in these Inferior Bodies. Some in Stones, some in Plants, some in Joints and Knots of Trees and their Boughs and some in various Members of Animals.

For the Bay Tree, Lote Tree and Marigold are Solary Herbs and their Roots and Knots being cut, they show the Characters of the sun. And in Stones the Character and Images of Celestial Things are often found. But there being so great a Diversity of Things, there is only a Traditional Knowledge of a Few Things which Human Understanding is able to reach.

Therefore, very few of those things are known to us, which the Ancient Philosophers and Chiromancers attained to, party by reason and partly by experience. And there yet lie hid may things in the Treasury of Nature, which the diligent student and wise searcher shall contemplate and discover.

Chapter 9
Treating of the Virtue and Efficacy of Perfumes
or Suffumigations and Vapors
and to what Plants they are properly and rightly attributed

It is necessary, before we come to the Operative or Practical Part of Talismanic Magic, to show the compositions of fumes or vapors, that are proper to the Stars and are of great force for the opportunely receiving of Celestial Gifts, under the Rays of the Stars.

Inasmuch as they strongly work upon the Air and Breath, for our Breath is very much changed by such Kind of Vapors, if both Vapors are of the other like; the Air being also, through the said Vapors, easily moved, or infected with the Qualities of Inferiors, or Celestial (daily quickly penetrating our Breast and Vitals), does wonderfully reduce us to the like qualities.

Let no man wonder how great things, Suffumigations can do in the Air, especially when they shall, with Porphyry, consider that, by certain Vapors exhaled from proper Suffumigations, Aerial Spirits are raised. Also Thunder and Lightnings and the like, as the Liver of a Chameleon being burnt on the housetop, will raise Showers and Lightnings; the same effects has the Head and Throat, if they are burnt with Oaken Wood.

There are some Suffumigations under the Influence of the Stars, that cause Images of Spirits to appear in the Air, or elsewhere. For if coriander, Smallage (Wild Celery), Henbane and Hemlock be made to fume, by Invocations, Spirits will soon come together, being attracted by the Vapors, which are most congruous to their own Natures. Hence, they are called the Herbs of the Spirits.

Also it is said, that if a Fume is made of the Root of the Reedy Herb Sagapen -- (thought to be mythical, but found on a medicinal herb website as, Latin name of, Ferula Szowitziana of the Family Umbelliferae, found in Turkey; see Natural Medicinal Herbs website online for the amplified reference) – with the Juice of Hemlock and Henbane (Poisonous Nightshade) and the Herbs, Tapsus Barbatus, Red Sanders and Black Poppy, it will likewise make strange shapes appear. But if a Suffume be made of Smallage, it chases them away and destroys their Visions.

Again, if a Perfume is made of Calamint, Peony, Mint and Palma Christi, it drives away ALL EVIL SPIRITS and Vain Imaginations.

Likewise, by certain Fumes, Animals are gathered together and put to flight.

Pliny mentions concerning the Stone, Liparis, that, with the Fume thereof, all Beasts are attracted together.

The Bones in the Upper Part of the Throat of a Hart, being burnt, bring Serpents together. But the Horn of the Hart, being burnt, chases away the same. Likewise, a Fume of Peacock's Feathers does the same. Also, the Lungs of an Ass, being burnt, puts all Poisonous Things to flight. And the Fume of the Burnt Hoof of a Horse, drives away Mice. The same does the Hoof of a Mule and with the Hoof of the Left Foot flies are driven away.

And if a house or ally place, is smoked with the Gall of Cuttle Fish, made into a confection with Red Storax, Roses and Lignum Aloes and then there is some Seawater or Blood cast into that place, the whole house will seem to be full of Water or Blood.

Now such Kind of Vapors as these, we must conceive, do infect a Body and infuse a Virtue into it which continues long, even as the Poisonous Vapor of the Pestilence, being kept for 2 years in the Walls of a House, infects the Inhabitants. And, as the Contagion of Pest or Leprosy hid in a garment, will, long after, infect him that wears it.

Now there are certain Suffumigations used to almost all our Instruments of Magic (of which hereafter), such as Images, Rings, etcetera. For some of the Magicians say, that if anyone shall hide Gold or Silver or any other such like Precious Thing (the Moon being in conjunction with the Sun) and shall Perfume the place with Coriander, Saffron, Henbane, Smallage and Black Poppy, of each the same quantity and bruised together and tempered with the Juice of Hemlock, that thing which is so hid shall never be taken away therefrom, but that Spirit shall continually keep it.

And if anyone shall endeavor to take it away by force, they shall be hurt or struck with a frenzy. And Hermes says, there is nothing like the Fume of Spermaceti for the raising up of Spirits. Therefore, if a Fume is made of that, Lignum Aloes, Pepperwort, Musk, Saffron and Red Storax, tempered together with the Blood of a Lapwing or Bat, it will quickly gather Airy Spirits to the place where it is used. And if it be used above the Graves of the Dead, it will attract Spirits and Ghosts thither.

Now the use of Suffumigations is this. That whenever we set about making any Talisman, Image or the like, under the Rule of Dominion of any Star or Planet, we should by no means omit the making of a Suffumigations appropriate to that Planet or Constellation under which we desire to work any effect or wonderful operation. As for instance; when we direct any work to the Sun, we must Suffume with Solary Things. If (we Suffume) to the Moon, we must do it with Lunary things. And so the rest.

And we must be careful to observe, that as there is a contrariety or Antipathy in the Natures of the Stars and Planets and their Spirits, so there is also in Suffumigations. For there is an Antipathy between Lignum Aloes and Sulphur Frankincense and Quicksilver. And Spirits that are raised by the Fume of Lignum Aloes are laid by the burning of Sulphur.

For the Learned Proclus gives an example of a Spirit that appeared in the Form of a Lion, furious and raging, by setting a White Cock before the apparition it soon vanished away. Because there is so great a contrariety between a Cock and a Lion. And let this suffice for a General Observation in these kind of things. We shall proceed with showing distinctly the Composition of the Several Fumes appropriated to the Seven Planets.

Chapter 10
Of the Composition of Some Perfumes
Appropriated to the Seven Planets

The SUN ~ ☉

We make a Suffumigation for the Sun in this Manner.

Take of Saffron, Ambergris, Musk, Lignum aloes, Lignum Balsam, the Fruit of the Laurel, Cloves, Myrrh and Frankincense, of each a like quantity, all of which, being bruised and mixed together, so as to make a sweet odor, must be incorporated with the Brain of an Eagle or the Blood of a White Cock, after the manner of Pills or Troches.

The MOON ~ ☽

For the Moon, we make a Suffume of the Head of a Frog dried and the Eyes of a Bull, the Seed of White Poppies, Frankincense and Camphor, which must be incorporated with Menstruous Blood or the Blood of a Goose.

SATURN ~ ♄

For Saturn take the Seed of Black Poppies, Hanbane, Mandrake Root, Loadstone and Myrrh and mix them up with the Brain of a Cat and the Blood of a Bat.

JUPITER ~ ♃

Take the Seed of Ash, Lignum Aloes, Storax, the Gum Benjamin, the Lapis Lazuli, the Tops of Peacocks' Feathers and incorporate with the Blood of a Stork, or Swallow or the Brain of a Hart.

MARS ~ ♂

Take Uphorbium, Bdellium, Gum Armoniac, the Roots of Both Hellebores and Loadstone and a Little Sulphur and incorporate them altogether with the Brain of a hart, the Blood of a Man and the Blood of a Black Cat.

VENUS ~ ♀

Take Musk, Ambergris, Lignum Aloes, Red Roses and Red Coral and make them up with Sparrow's Brains and Pigeon's Blood.

MERCURY ~ ☿

Take Mastich, Frankincense, Cloves and the Herb Cinquefoil and the Agate Stone and incorporate them all with the Brain of a Fox or Weasel and the Blood of a Magpie.

GENERAL FUMES OF THE PLANETS

To Saturn, are appropriated for Fumes, Odoriferous Roots, as Pepperwort Root, etcetera and the Frankincense Tree.

To Jupiter, are appropriated for Fruits, as Nutmegs, Cloves, etcetera.

To Mars, are appropriated, all Odoriferous Woods, as Sanders, Cyprus, Lignum Balsam and Lignum Aloes.

To the Sun, all Gums are appropriated, as Frankincense, Mastich Benjamin, Storax, Laudanum, Ambergris and Musk.

To Venus, Flowers are appropriated, as Roses, Violets, Saffron and the like.

To Mercury, are appropriated, all the Parings of wood or Fruit, as Cinnamon, Lignum Cassia, Mace, Citron Peel and Bayberries and whatever Seeds are Odoriferous.

To the Moon, are appropriated, the Leaves of all Vegetables, as the Leaf Indium, the Leaf of the Myrtle and Bay Tree.

Know also, that according to the opinion of all Magicians, in every good matter (as Love, Goodwill, etcetera), there must be a Good Perfume, Odoriferous and Precious. And in evil matters (as Hatred, Anger, Misery and the like), there must be a Stinking Fume that is of no worth.

The 12 Signs of the Zodiac also have their Proper Suffumigations, viz. Aries, Myrrh; Taurus, Pepperwort; Gemini, Mastich; Cancer, Camphor; Leo, Frankincense; Virgo, Sanders; Libra, Galbanum; Scorpio, Oppoponax; Sagittarius, Lignum Aloes; Capricorn, Benjamin; Aquarius, Euphorbium; Pisces, Red Storax.

But Hermes describes the most powerful Fume to be, that which is compounded of the Seven Aromatics, according to the Powers of the Seven Planets. For it receives from Saturn, Pepperwort; from Jupiter, Nutmeg; from Mars, Lignum Aloes; from the Sun, Mastich; from Venus, Saffron; from Mercury, Cinnamon; and from the Moon, Myrtle.

By a close observation of the above Order of Suffumigations, conjoined with other things, of which we shall speak hereafter (necessary to the full accomplishment of Talismanic Magic), many wonderful effects may be caused, especially if we keep in eye what was delivered in the First Part of Our Magic, viz. that the Soul of the Operator must go along with this.

Otherwise, in vain is Suffumigation, Seal, Ring, Image, Picture, Glass or any other Instrument of Magic; seeing that, it is not merely the disposition, but the ACT of disposition and firm and powerful INTENT or IMAGINATION that gives the effect.

We shall now hasten to speak, generally, of the Construction of Magical Rings and their wonderful and Potent Virtues and Operations.

Chapter II
Of the Composition and Magic Virtue of Rings

Rings, when they are opportunely made, impress their Virtues upon us insomuch that they affect the Spirit of him that carries them with gladness or sadness and render him bold or fearful, courteous or terrible, amiable or hateful.

Inasmuch, also as they fortify us against Sickness, Poisons, Enemies, Evil Spirits and All Manner of Hurtful Things and often, where the Law has no effect, these little trifles greatly assist and corroborate the troubled Spirit of the Wearer and help him, in a wonderful manner, to overcome his adversaries, while they do wonder how it is that they cannot effect any hurtful undertaking against him.

These things, I say, are great helps against Wrathful, Vicious, Worldly-Minded Men, inasmuch as they do terrify, hurt and render invalid the machinations of those who would otherwise work our Misery and Destruction.

All which we are neither afraid nor ashamed to declare, well knowing that these things will be hid from the Wicked and Profane, so as that they cannot draw the same into any abuse, or privy mischief toward their neighbor. We have reserved some few things in this Art to ourselves – not willing to throw Pearls before Swine.

And, however simple and plain we may describe some certain experiments and operations (so as that the great-mouthed School Philosophers may mutter or scoff thereat), yet there is nothing delivered in this Book but what may be, by an understanding thereof, brought into effect, and, likewise, out of which some good may be derived. But to proceed.

The Manner of Making these Rings is thus.

When any Star ascends in the Horoscope (fortunately), with a Fortunate aspect or conjunction of the Moon, we proceed to take a Stone and Herb, that is under that Star and likewise make a Ring of the Metal that is corresponding to the Star. And in the Ring under the Stone, put the Herb or Root, not forgetting to inscribe the Effect, Image, Name and Character, as also the Proper Suffume.

But I shall speak more of these in another place, where I speak of Images and Characters. Therefore, in making of Magical Rings, these things are unerringly to be observed as we have ordered. If anyone is willing to work any effect or experiment in Magic, he must by no means neglect the necessary circumstances which we have so uniformly delivered.

I have read, in Philostratus Jarchus, that a Prince of the Indians bestowed Seven Rings, marked with the Virtues and names of the Seven Planets, to Appollonius, of which he wore one every day, distinguishing according to the Names of the Days; by the benefit of which, he lived above 130 years, as also always retained the Beauty of his Youth.

In like manner, Moses, the Lawgiver and Ruler of the Hebrews, being skilled in the Egyptian Magic, is said, by Josephus, to have made Rings of Love and Oblivion. There was also, as says Aristotle, among the Cyreneans, a Ring of Battas, which could procure Love and Honor. We read, also, that Eudamus, a certain Philosopher, made Rings against the Bites of Serpents, Bewitchings and Evil Spirits. The same does Josephus relate of Solomon.

Also we read, in Plato, that Gygus, King of Lydia, had a Ring of Wonderful and Strange Virtues, the Seal of which, when he turned it toward the Palm of his Hand, nobody could see him, but he could see all things. By the opportunity of which ring, he ravished the Queen and slew the King, his Master and killed whomsoever he thought stood in his way. And in these Villainies, nobody could see him. And at length, by the Benefit of this Ring, he became King of Lydia.

Chapter 12
That the Passions of the Mind are Assisted by Celestials and that Constancy of Mind is in Every Work Necessary

The Passions of the Mind are much helped and are helpful and become most powerful, by Virtue of the Heaven, as they agree with the Heaven – either by any natural agreement or voluntary election.

For, as Ptolemy says.

"He who chooses that which is the better, seems to differ nothing from him who has this of Nature. It conduces, therefore, very much for the receiving the benefit of the Heavens, in any work, if we shall, by the Heaven, make ourselves suitable to it in our thoughts, affections, imaginations, elections, deliberations, contemplations and the like. For such like Passions vehemently stir up our Spirit to their likeness and suddenly expose us and ours to the Superior Significators of such like Passions. And also, by Reason of their Dignity and nearness to the Superiors, do partake more of the Celestials than any Material Things. For our mind can, through Imaginations or Reason, by a Kind of Imitation, be so conformed to any Star, as suddenly to be filled with the Virtues of that Star, as if we were a proper receptacle of the influence thereof.

Now the contemplating mind, as it withdraws itself from all sense, imagination, nature and deliberation and calls itself back to things separated, effects diverse things by Faith, which is a firm adhesion, a fixed intention and vehement application of the worker of receiver to him that co-operates in anything and gives power to the work which we intend to do. So that there is made, as it were, in us the Image of the Virtue to be received and the thing to be done in us or by us. We must, therefore, in every work and application of things, affect vehemently, imagine, hope and believe strongly, for that will be a great help. And it is verified amongst Physicians, that a strong belief and an undoubted hope and love towards the Physician, conduce much to Health, even more sometimes than the Medicine itself.

For the same that the Efficacy and Virtue of the Physician, by that means disposing himself for the receiving of the Virtue of the Physician and Physic. Therefore, he that works in Magic must be of a constant belief, be credulous and not at all doubt of the obtaining of the effect, for as a firm and strong belief does work wonderful things, although it be in false works, so distrust and doubting does dissipate and break the Virtue of the Mind of the worker, which is the Medium between both extremes. Whence it appears that he is frustrated of the desired influence of the Superiors, which could not be enjoined and united to our labors without a firm and solid Virtue of our mind.

Chapter 13
How Man's Mind may be joined with the Mind of Intelligences and Celestials and together with them, impress certain Wonderful Virtues upon Inferior Things

The Philosophers, especially the Arabians, say, that Man's Mind, when it is most intent upon any work, through its Passion and Effects, is joined with the Mind of the Stars and Intelligence, and, being so joined, is the cause that some Wonderful Virtue be infused into our works and things. And this, as because there is in it an Apprehension and Power of All Things, so because All Things have a Natural Obedience to it and of necessity, an efficacy, and more to that which desired them with a strong desire.

And according to this is verified the Art of Characters, Images, Enchantments and some speeches and many other wonderful experiments, to everything which the Mind affects.

By this means, whatsoever the Mind of him that is in vehement Love affects, has an efficacy to cause Love and whatsoever the Mind of him that strongly Hates, dictates, has an efficacy to hurt and Destroy.

The like is in Other Things which the Mind affects with a Strong Desire, for All Those Things which the Mind acts and dictates by Characters, Figures, Words, Speeches, Gestures and the like, help the Appetite of the Soul and acquire, Certain Wonderful Virtues, from the Soul of the Operator, in that hour when such a like Appetite does invade it, so from the opportunity and Celestial Influence, moving the Mind in this or that manner.

For Our Mind, when it is carried upon the Great Excess of any Passion or Virtue, oftentimes takes to itself a strong, better and more convenient hour or opportunity, which, Thomas Aquinas, in his Third Book Against the Gentiles, allows.

So, many Wonderful Virtues both cause and follow certain admirable operations by great affections, in those things which the Soul does dictate in that hour to them.

But know, that such Kind of Things confer nothing, or very little, but to the Author of them and to him who is inclined to them, as if he were the Author of them; and this is the Manner by which their efficacy is found out. And it is a general rule in them, that Every Mind, that is more excellent in its Desire and Affection, makes such Like Things more fit for itself, as also efficacious to that which it desires.

Everyone, therefore, that is willing to work in Magic, must know the Virtue, Measure, Order and Degree of his own Soul in the Power of the Universe.

Chapter 14
Showing the Necessity of Mathematical Knowledge and of the Great Power and Efficacy of Numbers in the Construction of TALISMANS, etcetera

The Doctrines of Mathematics are so necessary to and have such an Affinity with Magic, that they who profess it without them are quite out of the way and labor in vain and shall in no wise obtain their desired effect. For whatsoever things are and are done in these Inferior Natural Virtues, are all done and governed by Number, Weight, Measure, Harmony, Motion and Light.

And All Things which we see in these Inferiors have Root and Foundation in them, yet, nevertheless, without Natural Virtues of Mathematical Doctrines, only works like to Naturals can be produced.

As Plato says.

"A Thing not partaking of Truth or Divinity, but certain Images akin to them, as bodies going or speaking, which yet want the Animal faculty), such as were those which, amongst the Ancients, were called Daedalus' Images and (_) or which Aristotle makes mention, viz. the three-footed Images of Vulcan and Daedalus moving themselves; which Homer says, "came out of their own accord to the exercise."

And which, we read, moved themselves at the Feast of Hiaraba, the Philosophic Exerciser.

So there are made Glasses (some concave, other of the form of a Column) making the representation of Things in the Air seem like Shadows at a distance; of which sort Apollonius and Vitellius, in their Books, De Prospective, and (the other), Speculis, taught the making and the use.

And we read that Magnus Pompeius brought a certain Glass, amongst the spoils, from the East to Rome, in which were seen Armies of Armed Men. And there are made certain Transparent Glasses, which (being dipped in some certain Juices of Herbs and irradiated with an Artificial Light) fill the Whole Air Roundabout with Visions.

And we know how to make Reciprocal Glasses, in which the Sun Shining, All Things which were illustrated by the Rays thereof are apparently seen many miles off.

Hence a Magician (expert in Natural Philosophy and Mathematics and knowing the Middle Sciences, consisting of both these, viz. Arithmetic, Music, Geometry, Optics, Astronomy and such Sciences that are of Weights, Measures, Proportions, Articles and Joints. Knowing also, Mechanical Arts resulting from these, may, without any wonder, if he excel other men in that Art and Wit, do many Wonderful Things, which men may much admire.

There are some Relics now extant of the Ancients, viz. Hercules and Alexander's Pillars, the Gate of Caspia, made of Brass and shut with Iron Beams, that it could by no Art be broken. And the Pyramids of Julius Caeser, erected at Rome, near the Hill Vaticanus and the Mountains built by Art in the Middle of the Sea; and Towers and Heaps of Stones, such as I have seen in England, put together by Incredible Art.

But the Vulgar, seeing any Wonderful Sight, impute it to the Devil as his work; or think (it) a Miracle, which, indeed, is a Work of Natural or Mathematical Philosophy.

But here it is convenient that you know, that, as by Natural Virtues we collect Natural Virtues, so by Abstracted, Mathematical and Celestial, we receive Celestial Virtues, as Motion, Sense, Life, Speech, Soothsaying and Divination even in Matter less disposed, as that which is not made by Nature, but only by Art.

And so Images that speak and foretell Things to Come, are said to be made; as William of Paris relates of a Brazen Head, made under the Rising of Saturn, which, they say, spoke with a Man's Voice. But he that will choose a disposed Matter and most fit to receive and a most Powerful Agent, shall undoubtedly produce more Powerful Effects. For it is a General Opinion of the Pythagoreans, that, as Mathematical are more Formal than Natural, so also they are more efficacious; as they have less dependence in their Being, so also in their Operation.

But amongst all Mathematical Things, Numbers, as they have more of Form in them, so also are more efficacious, as well to affect what is Good as what is Bad.

All Things, which were first made by the Nature of Things in its First Age, seem to be formed by the Proportion of Numbers; for this was the Principal Pattern in the Mind of the Creator.

Hence is borrowed the Number of the Elements; hence the Courses of Times; hence the Motion of the Stars and the Revolution of the Heavens; and the State of All Things subsist by the Uniting Together of Numbers.

Numbers, therefore, are endowed with Great and Sublime Virtues. For it is no Wonder, seeing there are so many Occult Virtues in Natural Things, although of Manifest Operations, that there should be in Numbers much greater and more OCCULT and also more wonderful and efficacious. For as much as they are more formal, more perfect and naturally in the Celestials, not mixed with separated substances; and lastly, having the greatest and most simple commixing with the Ideas in the Mind of God, from which they receive their proper and most efficacious VIRTUES; wherefore, they also are of most force and conduce most to the obtaining of Spiritual and Divine Gifts – as, in Naturally Things, Elementary Qualities are Powerful in Transmuting of any Elementary Thing.

Again, All Things that are and are made, subsist by and receive their Virtue from Numbers; for Time consists of Numbers and All Motion and Action and All Things which are subject to Time and Motion.

Harmony, also and Voices, have their Power by and consist of Numbers and their Proportions; and the Proportion arising from Numbers do, by Lines and Points, make Characters and Figures; and these are Proper to Magical Operations – the Middle, which is between both, being appropriated by declining to the extremes, as in the Use of Letters. And lastly, All Species of Natural Things and of those which are above Nature, are joined together by Certain Numbers, which Pythagoras seeing, says, that Number is that by which All Things subsist and distributes each Virtue to each Number.

And Proclus says, Number has always a Being, yet there is One in Voice, another in Proportion of them, another in the Soul and Reason and another in Divine Things.

But Themistius, Boetius and Averrois the Babylonian, together with Plato, do so extol Numbers that they think no Man can be a True Philosopher without them. By them there is a way made for the searching out and understanding of All Things Knowable, by them the next access to Natural Prophesy is had; and the Abbott Joachim proceeded no other way in his Prophesies, but by Formal Numbers.

Chapter 15
The Great Virtues of Numbers as well in Natural Things as in Supernatural

That there lies wonderful efficacy and Virtue in Numbers, as well to Good as to Bad, the most Eminent Philosophers unanimously teach, especially Jerome (Hierom), Augustine (Austin), Origen, Ambrose, Gregory of Nazianzen, Athanasius, Basilius, Hilarius, Rubanas, Bede and many more conform.

Hence Hilarius, in his Commentaries upon the Psalms, testifies that the 70 Elders, according to the Efficacy of Numbers, brought the Psalms into order.

The Natural Number is not here considered. But the Formal consideration that is in the Number; and let that which we spoke before always be kept in mind, viz. that these Powers are not in Vocal Numbers of Merchants buying and selling, but in Rational, Formal and Natural. There are the distinct Mysteries of God and Nature.

But he who knows how to join together the Vocal Numbers and Natural with Divine and Order them into the same Harmony, shall be able to work and know Wonderful Things by Numbers, in which, unless there was a Great Mystery, John had not said in the REVELATION, "He that has understanding, let him compute the Number of the Name of the Beast, which is the Number of a Man."

And this is the most famous manner of computing amongst the Hebrews and Cabalists, as we shall show afterwards. But this you must know, that Simple Numbers signify Divine Things (in) Numbers of Ten; Celestial Numbers (signify Divine Things in Numbers) of 100;

Terrestrial Numbers (signify Divine Things in Numbers) of 1000; (and all, of) those things that shall be in a Future Age.

Besides, seeing the Parts of the Mind are according to an Arithmetical Mediocrity, by Reason of the Identity, or Equality of Excess, coupled together, but the Body, whose parts differ in their Greatness is according to a Geometrical Mediocrity, compounded; but an Animal consists of both, viz. Soul and Body, according to that Mediocrity which is suitable to Harmony; hence it is that Numbers work very much upon the Soul, Figures upon the Body and Harmony upon the Whole Animal.

Chapter 16
Of the SCALE of UNITY

Now let us treat particularly of Numbers themselves. And because Number is nothing else but a repetition of Unity, let us first consider Unity itself.

For Unity does most simply go through every Number and is the Common Measure, Fountain and Original of all Numbers; (it) contains every Number joined together in itself by entirety; the Beginning of every multitude, always the same and unchangeable, whence, also, being multiplied into itself, produces nothing but itself.

It is Indivisible, void of all parts. Nothing is before one, nothing is after one and beyond it is nothing and all things which are, desire that one, because all things proceed from One.

And that all things may be the same, it is necessary that they partake of that one and as all things proceed of one into many things, so all things endeavor to return to that one, from which they proceeded. It is necessary that they should put off multitude.

One, therefore, is referred to the Most High God, who, seeing he is One and Innumerable, yet creates Innumerable Things of Himself and contains them within Himself.

There is, therefore, One God, one World of the One God, one Sun of the One World, also One Phoenix in the World, One King amongst Bees, One Leader amongst Flocks of Cattle, One Ruler amongst Herds of Beasts and Cranes follow One; and many other Animals honor Unity. Amongst the Members of the Body there is One Principal, by which all the rest are guided, whether it be the Head, or (as some will) the Heart.

There is One Element, overcoming and penetrating All Things, viz. Fire. There is One Thing created of God, the subject of all wondering which is in the Earth or in the Heaven; it is actually Animal, Vegetable and Mineral; everywhere found, known by few, called by none by its proper name, but covered with figures and riddles, without which neither Alchemy, nor Natural Magic can attain to their complete end or perfection.

From one Man, Adam, all men proceeded. From that One, all men became Mortal. From that One, Jesus Christ, they are regenerated. And, as says St. Paul, One Lord, One Faith, One Baptism, One God and Father of All (are) One.

Mediator between God and Man, One Most High Creator, who is over all, by all and in us all, for there is One Father, God, from whence all and we in Him, One Lord Jesus Christ, by whom all and we by him, One God Holy Ghost, into whom all and we into Him (are framed).

THE DIAGRAM OF THE SCALE OF UNITY OF THE NUMBER ONE

1. In the Exemplary World ~

2. Jod ~

3. One Divine Essence, the Fountain of All Virtues and Power, whose name is expressed with one most simple letter

---

1. In the Intellectual World ~

2. The Soul of the World ~

3. One Supreme Intelligence, the First Creature, the Fountain of Life

---

1. In the Celestial World ~

2. The Sun ~

3. One King of Stars, the Fountain of Life

---

1. In the Elemental World ~

2. The Philosopher's Stone ~

3. One Subject and Instrument of all Virtues, Natural and Supernatural

Diagram of the Scale of UNITY of the NUMBER ONE (continued)

1. In the Lesser World ~

2. The Heart ~

3. One First Living and Last Dying

---

1. In the Infernal World ~

2. Lucifer ~

3. One Prince of Rebellion, of Angels and Darkness

Chapter 17
Of the NUMBER TWO and Its Scale

    The FIRST NUMBER IS TWO, because it is the First Multitude, it can be measured by no Number besides Unity alone, the common measure of all Numbers. It is not compounded of Numbers, but of One Unity only. Neither is it called a Number uncompounded, but more properly not compounded.

    The Number Three, is called the First Number uncompounded. But the Number Two is the First Branch of Unity and the First Procreation and it is called the Number of Science and Memory and of Light and the Number of Man, who is called another and the Lesser World. It is also called the Number of Charity and of Mutual Love; of Marriage and Society, as it is said by the Lord, "Two shall be called One Flesh."

And Solomon says, "It is better that Two be together than One, for they have a benefit by their mutual society. If one shall fall, he shall be supported by the other. Woe to him that is alone, because, when he falls, he has not another to help him. And if Two sleep together, they shall warm one another. How shall One be hot alone? And if any prevail against him, Two resist him."

And it is called the Number of Wedlock and Sex.

For there are Two Sexes, Masculine and Feminine.

And Two Doves bring forth Two Eggs, out of the first of which is hatched the Male, out of the second of which, is hatched the Female.

It is also called The Middle, that is Capable, that is Good and Bad, partaking. And the Beginning of Division, of Multitude and Distinction. And (it) signifies Matter.

This is also, sometimes, the Number of Discord, of Confusion, of Misfortune and Uncleanness.

(The Number from) whence St. Jerome (Hierom) against Jovianus (Jovanus) says, "That therefore, it was not spoken in the 2nd Day of the Creation of the World, (the words), 'And God said that it was Good.'"

(This) because the Number of Two is Evil.

Hence also it was, that god commanded that all Unclean Animals should go into the Ark by couples, because, as I said, the Number of Two is a Number of Uncleanness.

Pythagoras, as Eusebius reports, said, that Unity was God and a Good Intellect; but that Duality was a Devil and an Evil Intellect, in which is a Material Multitude.

Wherefore, the Pythagoreans say, that Two is not a Number, but a certain confusion of Unities.

And Plutarch writes, that the Pythagoreans called Unity, Apollo. And Two, they call Strife and Boldness. And Three, they call Justice, which is the Highest Perfection and is not without many Mysteries.

Hence there were Two Tables of the Law in Sinai; Two Cherubims looking to the propitiatory in Moses; Two Olives dropping Oil in Zachariah; Two Natures in Christ, Divine and Human.

Hence Moses saw Two Appearances of God, viz. His Face and Back Parts; also, Two Testaments; Two Commands of Love; Two First Dignities; Two First People; Two Kinds of Spirits, Good and Bad; Two Intellectual Creatures, an Angel and Soul; Two Great Lights; Two Solstitia; Two Equinoctials; Two Poles; Two Elements, producing a Living Soul, viz. Earth and Water.

THE DIAGRAM OF THE SCALE OF THE NUMBER TWO

1. In the Exemplary World ~

2. Jah ~
   EL ~

3. The Names of God, expressed with Two Letters ~
   (those are the Two Letter of either Jah or El in Hebrew)

The Diagram of the Scale of the NUMBER TWO (continued)

---

1. In the Intellectual World ~

2. An Angel ~
   The Soul ~

3. Two Intelligible Substances

---

1. In the Celestial World ~

2. The Sun ~
3. The Moon ~

4. Two Great Lights

---

1. In the Elementary World ~

2. The Earth ~
   The Water ~

3. Two Elements producing a Living Soul

---

1. In the Lesser World ~

2. The Heart ~
   The Brain ~

3. Two Principal Seats of the Soul

The Diagram of the Scale of the NUMBER TWO (continued)

1. In the Inferior World ~

2. Beemoth, weeping ~
   Leviathan, gnashing of teeth ~

3. Two Chiefs of the Devils ~
   Two Things Christ Threatens to the Damned

Chapter 18
Of the NUMBER THREE and Its Scale

The NUMBER THREE, is an Uncompounded Number, a Holy Number, a Number of Perfection, a Most Powerful Number.

For there are Three Persons in God; there are Three Theological Virtues in Religion. Hence it is that this Number conduces to the Ceremonies of God and Religion, that by the Solemnity of which, Prayers and Sacrifices are Three Times repeated.

For Corporeal and Spiritual Things consist of Three Things, viz. Beginning, Middle and End. By Three, as Trismegistus says, the World is Perfected; Harmony, Necessity and Order, ie. a concurrence of causes (which many call Fate) and the Execution of them to the Fruition or Increase, or a Due Distribution of the Increase.

The whole Measure of Time is concluded in Three, viz. Past, Present and Future (To Come). All Magnitude is contained in Three; Line, Superfices and Body. Every Body consists of Three Intervals; Length, Breadth and Thickness.

Harmony contains Three consents in Time; Diapason, Hemiolion, Diatesseron. There are also Three Kinds of Souls; Vegetative, Sensitive and Intellectual. And as such, says the Prophet, God orders the World by Number, Weight and Measure.

And the Number Three is deputed to the Ideal Forms thereof, as the Number Two is the Pro-creating Matter and Unity to God, the Maker of It.

Magicians do constitute Three Princes of the World; Oromasis, Mithris, Araminis; that is, God, the Mind and the Spirit.

By the Three Square or Solid, the Three Numbers of Nine, of Things Produced, are distributed, viz. of the Super-Celestial, into Nine Orders of Intelligences; of Celestial, into Nine Orbs; of Inferiors, into Nine Kinds of Generable and Corruptible Things.

Lastly, into this Eternal Orb, viz. Twenty-Seven, All Musical Proportions are included; as Plato and Proclus do at large discourse.

And the Number Three, has, in a Harmony of Five, the Grace of the First Voice.

Also, in Intelligences, there are Three Hierarchies of Angelical Spirits. There are Three Powers of Intellectual Creatures; Memory, Mind and Will. There are Three Orders of the Blessed, viz. Martyrs, Confessors and Innocents. There are Three Quaternions of Celestial Signs, viz. Fixed, Moveable and Common.

As also of Houses, there are Three (Types), viz. Centers, Succeeding and Falling. There are, also, Three Faces and Heads in Every Sign and Three Lords of Each Triplicity.

There are Three Fortunes amongst the Planets. In the Infernal Crew, (there are) Three Judges, Three Furies, Three Headed Cerberus. We read also, of a Thrice Double Hecate.

(There are) Three Months of the Virgin Diana. (There are) Three Persons in the Super-Substantial Divinity. (There are) Three Times of Nature, Law and Grace. (There are) Three Theological Virtues, (that are) Faith, Hope and Charity.

Jonah was Three Days in the Whale's Belly and so many was Christ in the Grave (of Three Days of total gone before Resurrection.)

(It comes to mind and to bear at this point of the book, since it is convenient to the examples of Numbers, that the Metaphysical dispensation of Philosophy can be interpreted easily enough in this case of wonderful things and magic as basically scientific enough. This is on the promise of a principal of insubstantial (and not in-substantive) means; it would seem that grammar in itself returns into the plethora of data, concerning any science of magic and so erase themselves as communication devices, to the extent that the insubstantial idiom can never be conveyed or learned, except by some success of practice of the exact rules or their meanings to a cite of proof which is both ethical and true and thereby, lending itself to some internal, mental meaning, which is hopefully correct in its mental symbolism and not corrupt, but exact and mathematical; though in any human way, it could not be a complete ideal.

So that, the insubstantial principle of which I mean, is the number itself. It needs not to be whatever human logic should dictate it to be, but it needs to be in order to interact with human logic, exactly the creative principle which it is. So that the human meaning of the coincidence of the number 1, 2, or 3 appearing need not satisfy human logic to satisfy magic, with or without human appearance, but by the logic or intelligence inherent in its creation as an aesthetic reality (insubstantial of any human concept other than to recognize its physical form in either a written theory or a scale of communications), as to represent both human and aesthetic means of logic.

Therefore, for its transcendence as both, it achieves a magical domain which touches the logic of humanity and can make itself known by its own means and principles and activities thereof; and thereby, we see its rules in the alter-aesthetic world, to seem coincidental to the logic we understand as universal and controlling. But we are not aesthetic numbers, theories or intelligences of the same form.)

THE DIAGRAM OF THE SCALE OF THE NUMBER THREE

1. In the Original World ~

2. The Father ~

3. Adai, The Son ~

4. The Holy Ghost ~

5. The Name of God with Three Letters

---

1. In the Intellectual World ~

2. Supreme Innocents ~

3. Middle Martyrs ~

4. Lowest of All Confessors ~

5. Three Hierarchies of Angels ~ Three Degrees of the Blessed

The Diagram of the Scale of the NUMBER THREE (continued)

1. In the Celestial World ~

2. Moveable, Corner, Of the Day ~

3. Fixed, Succeeding, Nocturnal ~

4. Common, Falling, Partaking ~

5. Three Quaternions of Signs ~ Three Degrees of the Blessed

1. In the Elementary World ~

2. Simple ~

3. Compounded ~

4. Three Times Compounded ~

5. Three Degrees of Elements

The Diagram of the Scale of the NUMBER THREE (continued)

1. In the Lesser World ~

2. The Head, in which the Intellect Grows,
   Answering to the Intellectual World ~

3. The Breast, where is the Heart, The Seat of Life,
   Answering to the Celestial World ~

4. The Belly, where the Faculty of Generation IS,
   and the Genital Members,
   Answering to the Elemental World ~

5. Three Parts,
   Answering to the Three-Fold World

---

1. In the Infernal World ~

2. Alecto, Minos, Wicked ~

3. Megera, Acacus, Apostates ~

4. Ctesiphone, Rhadamantus, Infidels ~

5. Three Infernal Furies, Three Infernal Judges,
   Three Degrees of the Damned

Chapter 19
Of the NUMBER FOUR and Its Scale

The Pythagoreans call the NUMBER FOUR – TECTRACTIS – and prefer it before all the Virtues of Numbers, because it is the Foundation and Root of all other Numbers.

Whence, also, all Foundations, as well in Artificial Things, as Natural and Divine, are Foursquare, as well shall show afterwards. And it signifies solidity, which also is demonstrated by a Foursquare Figure, for the Number Four, is the First Foursquare Plane, which consists of Two Proportions, whereof the First is of One to Two, the Latter of Two to Four. And it proceeds by a Double Procession and Proportion, viz. of One to One and of Two to Two; Beginning at a Unity and Ending at a Quaternity. Which proportions differ in this; that according to Arithmetic, they are unequal to one another, but according to Geometry, are equal.

Therefore, a Foursquare is ascribed to God the Father and also contains the Mystery of the Whole Trinity. For by its single proportion, viz. by the First of One to One, the Unity of the Paternal Substance is signified, from which proceeds One Son, equal to Him – by the next procession, also simple, viz. of Two to Two, is signified (by the second procession), the Holy Ghost.

For both (so that); the Son be equal to the Father, by the First Procession; and the Holy Ghost be equal to both, by the Second Procession.

(In the First Procession of One to One, the Son and the Father alone are equal and are the Unity of Paternal Substance and so it has a single proportion. In the Second Procession of Two to Two, the Holy Ghost (by the Mystery of the Trinity) is introduced to the proportions both, to conclude four syllogisms made up of two part with two parties each. So that, the Holy Ghost is equal to the Father and Son in Unity together as One Identity. So in the first, there are two identities as one to one and are first and in the second, there are three identities but in the first part of the second, unified only by the fact of the first and in the second part of the second, a singular identity added to them; so that taken as individuals, they remain the identities of the Trinity; wherein the offices of the Father and Son, see a double effect of duty and procession with the occurrence of the procession of the Holy Ghost in their company as a Unity.)

Hence that Super-Excellent and Great Name of the Divine Trinity in God, is written with Four Letters, viz. JODI, HE and VAU.

HE, is where it is the aspiration HE, signifies the proceeding of the Spirit from both. For HE, being duplicated, terminated both syllables and the whole name, but is pronounced JOVA, as some will, whence that JOVE of the HEATHEN, which the Ancients did picture with FOUR Cars.

Whence the Number Four, is the FOUNTAIN and HEAD of the Whole, DIVINITY.

And the Pythagoreans call it the Perpetual Fountain of Nature. For there are FOUR DEGREES in the Scale of Nature, viz. ~ to be, to live, to be sensible, to understand.

There are FOUR MOTIONS in Nature, viz. ~ Ascendant, descendant, going forward, circular.

There are FOUR CORNERS in Heaven, viz. ~ rising, falling, the Middle of the Heaven, the bottom of it.

There are FOUR ELEMENTS under Heaven, viz. ~ fire, air, water and earth.

According to these, there are FOUR TRIPLICITIES in Heaven.

There are FOUR FIRST QUALITIES under Heaven, viz. ~ cold, heat, dryness and moisture.

From these are the FOUR HUMORS, viz. ~ blood, phlegm, choler (heat sensitivity, blood pressure, hypothermic reactions, capillary insensitivities), melancholy.

Also (there are) FOUR PARTS of the YEAR ~ which are the spring, summer, autumn and winter.

Also (there are) FOUR PARTS of the WIND ~ divided into eastern, western, northern and southern.

There are also FOUR RIVERS in PARADISE and so many Infernal.

Also, the NUMBER FOUR makes up all KNOWLEDGE. First, it fills up every simple progress of Numbers with FOUR TERMS, viz. with one, two, three and four, constituting the NUMBER TEN.

It fills up every difference of Numbers.

The First Even and containing the First Odd in it.

It has in music, Diatesseron – the Grace of the Fourth Voice.

Also, it contains the Instrument of Four Strings and a Pythagorean Diagram, whereby are found out first of all, Musical Tunes and all Harmony of Music. For Double, Treble, Four Times Double, One and a Half, One and a Third Part, a Concord of All, a Double Concord of All, of Five, of Four; and all Consonance, is limited within the Bounds of the NUMBER FOUR.

It does also contain the Whole of Mathematics in FOUR TERMS, viz. point, line, superficies and profundity.

It comprehends all Nature in FOUR TERMS, viz. substance, quality, quantity and motion.

Also, all Natural Philosophy, in which are the Seminary Virtues of Nature, the natural springing, the growing form and the composites.

Also, Metaphysics is comprehended in FOUR BOUNDS, viz. being, essence, virtue and action.

Moral Philosophy is comprehended with FOUR VIRTUES, viz. prudence, justice, fortitude and temperance. It has also the Power of Justice ~ hence a FOUR FOLD LAW, viz. of providence, from God; of fatality, from the Soul of the World; of Nature, from Heaven; of prudence, from Man.

There are also FOUR JUDICIARY POWERS in all things being, viz. the intellect, discipline, opinion and sense.

Also, there are FOUR RIVERS of Paradise.

FOUR GOSPELS, received from FOUR EVANGELISTS, throughout the Whole Church (are those of authority).

The Hebrews received the Chief Name of God written with FOUR LETTERS.

Also the Egyptians, Arabians, Persians, Magicians, Mahometans, Grecians, Tuscans and Latins, write the name of God with FOUR LETTERS, viz. thus ~ Thet, Alla, Sire, Orsi, Abdi, θεὸς, Esar, Deus. Hence, the Lacedemonians, were wont to painting Jupiter with FOUR WINGS.

Hence, also, in Orpheus' Divinity, it is said that Neptune's Chariots are drawn with Four Horses.

There are also FOUR KINDS of DIVINE FURIES proceeding from several DEITIES, viz. from the Muses, Dionysius, Apollo and Venus.

Also the Prophet Ezekiel saw FOUR BEASTS of the RIVER CHOBAR and FOUR CHERUBIMS in FOUR WHEELS.

Also in Daniel, FOUR GREAT BEASTS did ascend from the SEA and FOUR WINDS did fight.

And in the REVELATIONS, FOUR BEASTS were Full of EYES, before and behind, standing roundabout the THRONE OF GOD and FOUR ANGELS, to whom was given Power to Hurt the EARTH and the SEA, did stand upon the FOUR CORNERS Of the EARTH, holding the FOUR WINDS, that they should not blow upon the EARTH, nor upon the SEA, nor upon any TREE.

THE DIAGRAM OF THE SCALE OF THE NUMBER FOUR

1. The Name of God with Four Letters ~

   יהוה ~
   (corresponds to #2 - #5 in following chart sequences)

   In the Original World, whence the Law of Providence ~
   (corresponds to last un-numbered lines in subsequent sequences)

---

1. Four Triplicities, or Intelligible Hierarchies ~

2. Seraphim, Cherubim, Thrones ~

3. Dominations, Powers, Virtues ~

4. Principalities, Archangels, Angles ~

5. Innocents, Martyrs, Confessors ~

   In the Intellectual World, whence the Fatal Law ~

The Diagram of the Scale of the NUMBER FOUR (continued)

1. Four Angels ruling over the Four Corners of the World ~
2. Michael ~
3. Raphael ~
4. Gabriel ~
5. Uriel ~

1. Four Rulers of the Elements ~
2. Seraph ~
3. Cherub ~
4. Tharsis ~
5. Ariel ~

1. Four Consecrated Animals ~
2. The Lion ~
3. The Eagle ~
4. Man ~
5. Calf ~

The Diagram of the Scale of the NUMBER FOUR (continued)

1. Four Triplicities of the Tribes of Israel ~
2. Dan, Asser, Naphthalin ~
3. Jehuda, Isachar, Zebulun ~
4. Manasse, Benjamin, Ephraim ~
5. Reuben, Simeon, Gad ~

1. Four Triplicities of the Apostles ~
2. Matthias, Peter, Jacob the Elder ~
3. Simon, Bartholomew, Matthew ~
4. John, Philip, James the Younger ~
5. Thaddeus, Andrew, Thomas ~

The Diagram of the Scale of the NUMBER FOUR (continued)

1. Four Evangelists ~
2. Mark ~
3. John ~
4. Matthew ~
5. Luke ~

1. Four Triplicities of Signs ~
2. Aries, Leo, Sagittarius ~
3. Gemini, Libra, Aquarius ~
4. Cancer, Scorpion, Pisces ~
5. Taurus, Virgo, Capricornus ~

   In the Celestial World, where is the Law of Nature ~

The Diagram of the Scale of the NUMBER FOUR (continued)

1. The Stars and Planets related to the Elements ~
2. Mars and the Sun ~
3. Jupiter and Venus ~
4. Saturn and Mercury ~
5. The Fixed Stars and the Moon ~

---

1. Four Qualities of Celestial Elements ~
2. Light ~
3. Diaphanousness ~
4. Agility ~
5. Solidity ~

The Diagram of the Scale of the NUMBER FOUR (continued)

---

1. Four Elements ~

2. Fire ~

3. Air ~

4. Water ~

5. Earth ~

   In the Elementary, where the Law of Generation and Corruption is ~

---

1. Four Seasons ~

2. Summer ~

3. Spring ~

4. Winter ~

5. Autumn ~

   In the Elementary, where the Law of Generation and Corruption is ~

The Diagram of the Scale of the NUMBER FOUR (continued)

1. Four Corners of the World ~

2. East ~

3. West ~

4. North ~

5. South ~

   In the Elementary, where the Law of Generation and Corruption is ~

1. Four Perfect Kinds of Mixed Bodies ~

2. Animals ~

3. Plants ~

4. Metals ~

5. Stones ~

   In the Elementary, where the Law of Generation and Corruption is ~

The Diagram of the Scale of the NUMBER FOUR (continued)

1. Four Kinds of Animals ~

2. Walking ~

3. Flying ~

4. Swimming ~

5. Creeping ~

In the Elementary, where the Law of Generation and Corruption is ~

1. What Answers the Elements in Plants ~

2. Seeds ~

3. Flowers ~

4. Leaves ~

5. Roots ~

In the Elementary, where the Law of Generation and Corruption is ~

The Diagram of the Scale of the NUMBER FOUR (continued)

1. What Answers the Elements in Metals ~

2. Gold and Iron ~

3. Copper and Tin ~

4. Quick Silver ~

5. Lead and Silver ~

   In the Elementary, where the Law of Generation and Corruption is ~

1. What Answers the Elements in Stones ~

2. Bright and Burning ~

3. Light and Transparent ~

4. Clear and Congealed ~

5. Heavy and Dark ~

   In the Elementary, where the Law of Generation and Corruption is ~

The Diagram of the Scale of the NUMBER FOUR (continued)

1. Four Elements of Man ~

2. The Mind ~

3. Spirit ~

4. Soul ~

5. Body ~

   In the Lesser World, viz. Man from Whom is the Law of Prudence ~

1. Four Powers of the Soul ~

2. The Intellect ~

3. Reason ~

4. Fantasy ~

5. Sense ~

   In the Lesser World, viz. Man from Whom is the Law of Prudence ~

The Diagram of the Scale of the NUMBER FOUR (continued)

1. Four Judiciary Powers ~

2. Faith ~

3. Science ~

4. Opinion ~

5. Experience ~

   In the Lesser World, viz. Man from Whom is the Law of Prudence ~

1. Four Moral Virtues ~

2. Justice ~

3. Temperance ~

4. Prudence ~

5. Fortitude ~

   In the Lesser World, viz. Man from Whom is the Law of Prudence ~

The Diagram of the Scale of the NUMBER FOUR (continued)

1. The Senses Answering to the Elements ~

2. Sight ~

3. Hearing ~

4. Taste and Smell ~

5. Touch ~

In the Lesser World, viz. Man from Whom is the Law of Prudence ~

1. Four Elements of Man's Body ~

2. Spirit ~

3. Flesh ~

4. Humors ~

5. Bones ~

In the Lesser World, viz. Man from Whom is the Law of Prudence ~

The Diagram of the Scale of the NUMBER FOUR (continued)

---

1. A Four-Fold Spirit ~

2. Animal ~

3. Vital ~

4. Generative ~

5. Natural ~

    In the Lesser World, viz. Man from Whom is the Law of Prudence ~

---

1. Four Humors ~

2. Choler (as Passion) ~

3. Blood (as Vitality) ~

4. Phlegm (as Gall) ~

5. Melancholy (as Error) ~

    In the Lesser World, viz. Man from Whom is the Law of Prudence ~

The Diagram of the Scale of the NUMBER FOUR (continued)

---

1. Four Manners of Complexity (as Composition) ~

2. Violence ~

3. Nimbleness ~

4. Dullness ~

5. Slowness ~

   In the Lesser World, viz. Man from Whom is the Law of Prudence ~

---

1. Four Princes of Devils, Offensive in the Elements ~

2. Samael ~

3. Azazel ~

4. Azael ~

5. Mahazael ~

   In the Infernal World, where is the Law of Wrath and Punishment ~

The Diagram of the Scale of the NUMBER FOUR (continued)

1. Four Infernal Rivers ~

2. Phelgethon ~

3. Cocytus ~

4. Styx ~

5. Acheron ~

---

1. Four Princes of Spirits, Upon the Four Angles of the World ~

2. Oriens ~

3. Paymon ~

4. Egyn ~

5. Amaymon ~

Chapter 20
Of the NUMBER FIVE and Its Scale

The NUMBER FIVE is of no small force., for it consists of the FIRST EVEN and the FIRST ODD; of which, the Odd Number is the Male and the Even Number is the Female, whence Arithmetics call that the Father and this the Mother.

Therefore, the NUMBER FIVE is of no small Perfection or Virtue, which proceeds from the commixing of these Numbers.

It is, also, the just MIDDLE of the UNIVERSAL NUMBER, which is TEN; for if you sub-divide the NUMBER TEN into its possible whole un-fractioned portions, there will be NINE and ONE, or EIGHT and TWO, or SEVEN and THREE, or SIX and FOUR; and every collection makes the NUMBER TEN and the EXACT MIDDLE is always the NUMBER FIVE and its equidistant.

And therefore, the NUMBER FIVE is called by the Pythagoreans, the NUMBER of WEDLOCK, as also of JUSTICE, because it divides the NUMBER TEN in an Even Scale.

There are FIVE SENSES in Man; sight, hearing, smelling, tasting and feeling.

There are FIVE POWERS in the Soul; vegetative, sensitive, concupiscible, irascible and rational.

There are FIVE FINGERS on the Hand.

There are FIVE WANDERING PLANETS in the Heavens, according to which there are FIVE-FOLD TERMS in every Sign.

In Elements, there are FIVE KINDS of MIXED BODIES, viz. stones, metals, plants, plant-animals, animals and so many Kinds of

Animals; as men, four-footed beasts, creeping, swimming and flying.

And there are FIVE KINDS by which all things are made of God, viz. essence, the same, another, sense and motion.

The Swallow brings forth but FIVE YOUNG, which she feeds with equity, beginning with the eldest and so the rest, according to their age. For in this Number, the Father Noah found favor with God and was preserved in the Flood of Waters.

In the Virtue of this Number, Abraham, being 100 years old, begat a Son of Sarah (Sarah being 90 years old and a barren woman and past child-bearing) and grew up to be a Great People.

Hence in Time of Grace, the name of Divine Omnipotent is called upon in FIVE LETTERS.

In Time of Nature, the name of God was called upon with THREE LETTERS, Sadai or שדי.

In Time of the Law, the ineffable name of God was expressed with FOUR LETTERS ~ יהוה ~ instead of which the Hebrews express ~ אדני ~ Adonai.

In Time of Grace, the ineffable name of God was written with FIVE LETTERS ~ יהשוה, ~ Jhesu, which is called upon with no less mystery than that of THREE LETTERS ~ ישך.

DIAGRAM OF THE SCALE OF THE NUMBER FIVE

1. The Names of God with Five Letters ~
   The Name of Christ with Five Letters ~

4. Eloim ~
   Elohi ~
   Jhesu ~

7. In the Exemplary World

---

1. Five Intelligible Substances ~

2. Spirits of the First Hierarchy, called Gods, or the Sons of Gods ~

3. Spirits of the Second Hierarchy, called Angels which are Sent ~

4. Spirits of the Third Hierarchy, called Angels which are Sent ~

5. Souls of Celestial Bodies ~

6. Heroes and Blessed Souls ~

7. In the Intellectual World ~

Diagram of the Scale of the NUMBER FIVE (continued)

1. Five Wandering Stars, Lords or the Terms ~
2. Saturn ~
3. Jupiter ~
4. Mars ~
5. Venus ~
6. Mercury ~
7. In the Celestial World ~

1. Five Kinds of Corruptible Things ~
2. Water ~
3. Air ~
4. Fire ~
5. Earth ~
6. A Mixed Body ~
7. In the Elementary World ~

Diagram of the Scale of the NUMBER FIVE (continued)

---

1. Five Kinds of Mixed Bodies ~
2. Animal ~
3. Plant ~
4. Metal ~
5. Stone ~
6. Plant-Animal ~
7. (In the Universal, Bestial and/or Mutable World ~)

---

1. Five Senses ~
2. Taste ~
3. Hearing ~
4. Seeing ~
5. Touching ~
6. Smelling ~
7. In the Lesser World ~

Diagram of the Scale of the NUMBER FIVE (continued)

1. Five Corporeal Torments ~

2. Deadly Bitterness ~

3. Horrible Howling ~

4. Terrible Darkness ~

5. Unquenchable Heat ~

6. A Piercing Stink ~

7. In the Infernal World ~

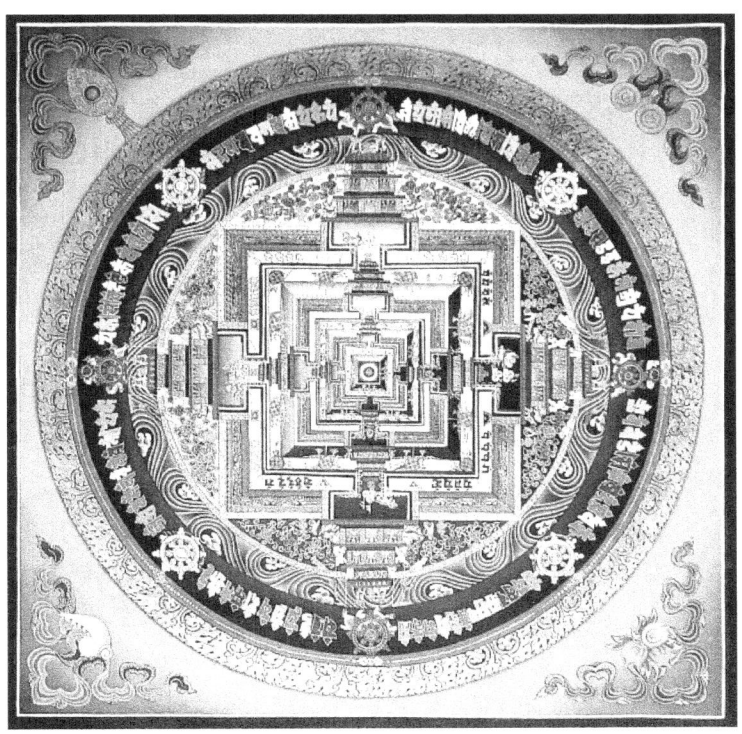

Chapter 21
Of the NUMBER SIX and Its Scale

SIX is a NUMBER of PERFECTION, because it is the Most Perfect in NATURE, in the Whole Course of NUMBERS, from ONE to TEN. And it alone is so PERFECT that in the Collection of its parts, it results the same, neither wanting nor abounding.

For if the Parts thereof, viz. the MIDDLE, THIRD and SIXTH PART, which are THREE, TWO, ONE, be gathered together, they perfectly fill up the Whole Body of SIX, which Perfection all the Other NUMBERS want.

Hence, by the Pythagoreans, it is said to be altogether to be applied to Generation and Marriage and is called the SCALE of the WORLD.

For the WORLD is made of the NUMBER SIX.

Neither does it abound, nor is (it) defective.

Hence that is, because the WORLD was finished by GOD, the SIXTH DAY (it is represented by the NUMBER SIX).

For the SIXTH DAY, GOD saw all things which HE had made and they were VERY GOOD.

Therefore, the HEAVEN and the EARTH and all the HOST thereof, were finished.

It is also called the NUMBER of MAN, because the SIXTH DAY, MAN was Created.

And it is also the NUMBER of OUR REDEMPTION, for on the SIXTH DAY, CHRIST suffered for OUR REDEMPTION, whence there is a Great Affinity between the NUMBER SIX and the CROSS, Labor and Servitude.

Hence, it is commanded in the LAW, that in SIX DAYS, the MANNA is to be gathered and WORK to be done.

SIX YEARS, the GROUND was to be SOWN.

And (also) the HEBREW SERVANT was to SERVE his MASTER SIX YEARS.

SIX DAYS the GLORY of the LORD appeared upon MOUNT SINAI, covering it with a CLOUD.

The CHERUBIMS had SIX WINGS.

(There are) SIX CIRCLES in the FIRMAMENT ~ ARTIC, ANTARCTIC, TWO TROPICS, EQUINOCTIAL and ECLIPTICAL.

(There are) SIX WANDERING PLANETS ~ SATURN, JUPITER, MARS, VENUS, MERCURY, the MOON ~ running through the LATITUDE of the ZODIAC on both sides of the ECLIPTIC.

There are SIX SUBSTANTIAL QUALITIES in the ELEMENTS, viz. sharpness, thinness, motion; (with) the Contrary to these (which are) ~ dullness, thickness and rest.

There are SIX DIFFERENCES of POSITION ~ upwards, downwards, before, behind, on the right side, and, on the left side.

There are SIX NATURAL OFFICES, without which, NOTHING can be, viz. magnitude, color, figure, interval, standing, motion.

Also, a SOLID FIGURE of any FOURSQUARE THING has SIX SUPERFICES.

There are SIX TONES of all HARMONY, viz. five tones and two half tones which make one tone, which is the SIXTH TONE.

DIAGRAM OF THE SCALE OF THE NUMBER SIX

1. In the Exemplary World ~

8. Names of Six Letters ~

---

1. In the Intelligible World ~

2. Seraphim ~

3. Cherubim ~

4. Thrones ~

5. Dominations ~

6. Powers ~

7. Virtues ~

8. Six Orders of Angels, which are not sent to Inferiors ~

Diagram of the Scale of the NUMBER SIX (continued)

1. In the Celestial World ~

2. Saturn ~
3. Jupiter ~
4. Mars ~
5. Venus ~
6. Mercury ~
7. The Moon ~

8. Six Planets wandering through the Latitude of the Zodiac from the Ecliptic ~

1. In the Elemental World ~

2. Rest ~
3. Thinness ~
4. Sharpness ~
5. Dullness ~
6. Thickness ~
7. Motion ~

8. Six Substantial Qualities of the Elements ~

Diagram of the Scale of the NUMBER FIVE (continued)

1. In the Lesser World ~

2. The Intellect ~
3. Memory ~
4. Sense ~
5. Motion ~
6. Life ~
7. Essence ~

8. Six Degrees of the Mind ~

1. In the Infernal World ~

2. Acteus ~
3. Megalesius ~
4. Ormenus ~
5. Lycus ~
6. Nicon ~
7. Mimon ~

8. Six Devils, the Authors of All Calamities ~

Chapter 22
Of the NUMBER SEVEN and Its Scale

The NUMBER SEVEN is of Various and Manifold Power, for it consists of ONE and SIX, or of TWO and FIVE, or of THREE and FOUR. And it has a UNITY, as it were the coupling together of TWO THREE's. Whence if we consider the several parts thereof and the joining together of them, without doubt, we shall confess that it is, as well by the joining together of the parts thereof, as by its fullness apart, most full of all Majesty.

And the Pythagoreans call it the VEHICLE of Man's Life, which is does not receive from its Part So; as it Perfects by its proper right of its Whole; for it contains BODY and SOUL.

For the Body consists of FOUR ELEMENTS and is endowed with FOUR QUALITIES, also, the NUMBER THREE respects the Soul, by reason of its THREE-FOLD Power, viz. rational, irascible and concupiscible.

The NUMBER SEVEN, therefore, because it consists of THREE and FOUR, joins the Soul to the Body and the Virtue of this NUMBER relates to that of men and it causes Man to be received, formed, brought forth, nourished, live and indeed altogether to subsist. For when the Genital Seed is received in the Womb of the Woman, if it remains there SEVEN HOURS after the effusion of it, it is certain that it will abide there for good.

Then the First SEVEN DAYS it is coagulated and is fit to receive the Shape of a Man. Then it produces Mature Infants, which are called INFANTS of the SEVENTH MONTH, that is, because they are born the SEVENTH MONTH (of their inborn infancy).

After the Birth, the SEVENTH HOUR tries whether it will live or no; for that which will bear the Breath of the Air after that hour, is conceived (and) will live.

After SEVEN DAYS, it casts off the Relics of the Navel. After Twice SEVEN DAYS, its Sight begins to move after the Light.

In the Third SEVEN DAYS, it turns its Eyes and Whole Face freely.

After SEVEN MONTHS, it breeds Teeth.

After the Second SEVEN MONTHS, it sits without Fear of Falling.

After the Third SEVEN MONTHS, it begins to speak.

After the Fourth SEVEN MONTHS, it stands strongly and walks.

After the Fifth SEVEN MONTHS, it begins to refrain sucking its Nurse.

After SEVEN YEARS, its First Teeth fall and new are bred, fitter for harder meat and its Speech is perfected.

After the Second SEVEN YEARS, boys wax ripe and then it is a Beginning of Generation.

At the Third SEVEN YEARS, they grow to men in stature and begin to be hairy and become able and strong for Generation.

At the Fourth SEVEN YEARS, they cease to Grow Taller.

In the Fifth SEVEN YEARS, they attain to the Perfection of their Strength.

(In) the Sixth SEVEN YEARS, they keep their Strength.

(In) the Seventh SEVEN YEARS, they attain to their utmost Discretion and Wisdom and the Perfect Age of Men.

But when they come to the Tenth SEVEN YEARS, where the NUMBER SEVEN is taken for a Complete Number, then they come to the Common Term of Life, to the Prophet saying, Our Age is in SEVENTY (70) YEARS.

The utmost height of a Man's Body is SEVEN FEET.

There are, also, SEVEN DEGREES in the Body, which complete the Dimension of its Altitude from the Bottom to the Top, viz. marrow, bone, nerve, vein, artery, flesh and skin.

There are SEVEN, which, by the Greeks, are called Black Members, which are ~ tongue, heart, lungs, liver, spleen and the two kidneys.

There are also, SEVEN PRINCIPAL PARTS of the Body, which are ~ head, breast, hands, feet and the privy members.

It is manifest, concerning Breath and Meat, that, without Drawing of Breath, the Life does not remain above SEVEN HOURS. And they that are Starved with Famine, live not above SEVEN DAYS.

The Veins, also, and Arteries, as Physicians say, are moved by the SEVENTH NUMBER.

Also, Judgments in Diseases are made with Greater Manifestation upon the SEVENTH DAY, which Physicians call Critical, that is, Judicial.

Also, of SEVEN PORTIONS, God creates the Soul. The Soul, also, receives the Body by SEVEN DEGREES.

All differences of Voices proceeds to the SEVENTH DEGREE, after which there is the same Revolution.

Again, there are SEVEN MODULATIONS of the Voices, that are ~ ditonus, semiditonus, Diatesseron, diapente with a tone, diapente with a half tone and diapason.

There are also, in Celestials, a most Potent Power of the NUMBER SEVEN, which is for seeing there are FOUR CORNERS of the HEAVEN diametrically looking one towards the other, which indeed is accounted a most full and powerful aspect and consists of the NUMBER SEVEN.

For it is made with the SEVENTH SIGN and makes a CROSS, the most Powerful Figure of ALL, which we shall speak in its due place.

But this you must not be ignorant of, that the NUMBER SEVEN has a Great Communion with the Cross.

By the same RADIATION and NUMBER, the SOLSTICE is distant from Winter and the Winter Equinox (is distant) from the Summer ~ all of which are accomplished with SEVEN SIGNS.

There are also SEVEN CIRCLES in the Heavens, according to the Longitudes of the Axle Tree.

There are SEVEN STARS about the Arctic Pole, Greater and Lesser, called Charles Wain.

Also, there are SEVEN STARS called the Pleiades.

And there are SEVEN PLANETS, according to those SEVEN DAYS constituting a Week.

The Moon is the SEVENTH of the PLANETS and next to us, observing this NUMBER more than the rest, this NUMBER dispensing the Motion and Light thereof.

For in 28 days, it runs around the Compass of the Whole Zodiac, which Number of Days, the NUMBER SEVEN with its SEVEN TERMS, viz. from ONE to SEVEN, does make and fill up as much as the SEVERAL NUMBERS, by adding to the antecedents and makes FOUR times SEVEN DAYS, in which the Moon runs through and about all the Longitude and Latitude of the ZODIAC, by measuring and measuring again.

With the like SEVEN DAYS, it dispenses its LIGHT, by changing it. For the First SEVEN DAYS, unto the Middle, as it were of the Divided World, it increases.

The Second SEVEN DAYS it fills its Whole Orb with LIGHT.

The Third SEVEN DAYS, it decreases its Orb of LIGHT and is again contracted into a Divided Orb.

But after the Fourth SEVEN DAYS, it is renewed with the last diminution of its LIGHT.

And by the Same SEVEN DAYS, it disposes the increase and decrease of the SEA, for in the First SEVEN DAYS of the increase of the Moon, it is by little and little lessened.

In the Second SEVEN DAYS, it is little by little increased.

But in the Third SEVEN DAYS, it is like the First and the Fourth SEVEN DAYS, does the same as the Second.

It is also applied to Saturn, which ascending from the lower, is the SEVENTH PLANET, which betokens rest; to which, the SEVENTH DAY is ascribed, which signifies the SEVEN 1000th, wherein, as St. John says, the Dragon (which is the Devil) and Satan being bound, men shall be quiet and lead a peaceable Life.

And the Leprous Person that was to be cleansed was sprinkled SEVEN TIMES with the Blood of a Sparrow.

And Elisha the Prophet, as it is written in the Second Book of Kings, says, unto the Leprous Person, "Go and wash yourself SEVEN TIMES in Jordan and your Flesh shall be made Whole and you shall be Cleansed."

Also, the NUMBER SEVEN is a NUMBER of REPENTANCE and REMISSION. And CHRIST, with SEVEN PETITIONS, finished his Speech of our satisfaction.

The NUMBER SEVEN is called the NUMBER of LIBERTY, because the SEVENTH YEAR, the Hebrew Servant did challenge Liberty for himself.

It is also most suitable to DIVINE PRAISES, whence the Prophet says, "SEVEN TIMES a day do I praise you, because of your Righteous Judgments."

It is moreover called the NUMBER of REVENGE, as says the Scripture, "And Cain shall be revenged SEVEN-FOLD."

And the Psalmist says, "Render unto Our Neighbors SEVEN-FOLD into their Bosom their reproach."

Hence there are SEVEN WICKEDNESSES, as says Solomon.

And there are SEVEN WICKED SPIRITS taken, as are read in the Gospel.

The NUMBER SEVEN signifies, also, the Time of the Present Circle, because it is finished in the Space of SEVEN DAYS.

Also it is consecrated to the HOLY GHOST, which the Prophet Isaiah describes to be SEVEN-FOLD, according to his Gift, viz. the Spirit of Wisdom and Understanding, the Spirit of Counsel and Strength, the Spirit of Knowledge and Holiness, the Spirit of Fear of the Lord, which we read in Zachariah to be the SEVEN EYES of GOD.

There are also SEVEN ANGELS, SPIRITS standing the PRESENCE of GOD, as is read in Tobias and in the REVELATION.

SEVEN LAMPS did burn before the THRONE of GOD and SEVEN GOLDEN CANDLESTICKS and in the Middle thereof was one like unto the SON of MAN and he had in his Right Hand, SEVEN STARS.

Also, there were SEVEN SPIRITS before the THRONE of GOD and SEVEN ANGELS stood before the THRONE of GOD and there were given to them, SEVEN TRUMPETS.

And he saw a LAMB, having SEVEN HORNS and SEVEN EYES and he saw the BOOK Sealed with SEVEN SEALS and when the SEVENTH SEAL was opened, there was Silence in Heaven.

Now, by all that has been said, it is apparent that the NUMBER SEVEN, amongst the other NUMBERS, may be deservedly said to be most Full of Efficacy.

Moreover, the NUMBER SEVEN has great conformity with the NUMBER TWELVE, for as THREE and FOUR make SEVEN, so THREE times FOUR makes TWELVE, which are the NUMBERS of the CELESTIAL PLANETS and SIGNS resulting from the same Root.

And by the NUMBER THREE partaking of the DIVINITY and by the NUMBER FOUR of the NATURE of INFERIOR THINGS.

There is in SACRED WRIT a very Great Observance of this NUMBER before all others and many and very great are the MYSTERIES thereof.

Many we have decreed to reckon up here, repeating them out of HOLY WRIT, by which it will easily appear that the NUMBER SEVEN does signify a Certain Fullness of Sacred Mysteries for we read, in Genesis, that the SEVENTH was the Day of Rest of the LORD, that Enoch, a Pious Holy Man, was the SEVENTH from ADAM.

And that there was another SEVENTH MAN from ADAM, a Wicked Man, by name of LAMECH, that had TWO Wives and that the SIN of CAIN, should be abolished the SEVENTH GENERATION, as it is written, "Cain shall be punished SEVEN-FOLD; and that he who shall slay Cain, shall be revenged SEVEN-FOLD."

To which, the Master of the History, collects that there were SEVEN SINS of CAIN.

Also, of all Clean Beasts ~ SEVEN and SEVEN were brought back into the ARK, as also of Fowls.

And after SEVEN DAYS, the LORD rained upon the EARTH and upon the SEVENTH Day the Fountains of the Deep were broken up and the WATERS covered the EARTH.

Also, ABRAHAM gave to ABIMELECH, SEVEN EWE LAMBS.

And, JACOB served SEVEN YEARS for LEAH and SEVEN more for RACHEL.

And SEVEN DAYS, the People of Israel bewailed the DEATH of JACOB.

Moreover, we read in the same place, of SEVEN KINE (Seven Kinds of Cattle) and SEVEN YEARS of CORN, SEVEN YEARS of PLENTY and SEVEN YEARS of SCARCITY.

And in EXODUS, the SABBATH of SABBATHS, the HOLY REST to the LORD, is commanded to be on the SEVENTH DAY.

Also, on the SEVENTH DAY, MOSES ceased to pray.

On the SEVENTH DAY, there shall be a Solemnity of the LORD.

The SEVENTH YEAR, the Servant shall go out FREE.

SEVEN DAYS let the CALF and the LAMB be with its DAM.

The SEVENTH YEAR, let the Ground that has been Sown SIX YEARS, be at REST.

The SEVENTH DAY shall be a HOLY SABBATH and a REST.

The SEVENTH DAY, because it is the SABBATH, shall be called HOLY.

In LEVITICUS, the SEVENTH DAY also shall be more observed and be More HOLY.

And the First Day of the SEVENTH MONTH shall be a SABBATH of MEMORIAL.

SEVEN DAYS shall the SACRIFICES be Offered to the LORD.

SEVEN DAYS shall the HOLY DAYS of the LORD be celebrated.

SEVEN DAYS in a YEAR EVERLASTING in the Generations.

In the SEVENTH MONTH you shall Celebrate FEASTS and shall dwell in TABERNACLES, SEVEN DAYS.

SEVEN TIMES he shall sprinkle himself before the LORD that has dipped his Finger in BLOOD.

He that is Cleansed from the LEPROSY, shall dip SEVEN TIMES in the BLOOD of a SPARROW.

SEVEN DAYS shall she be washed with running water, that is MENSTRUOUS.

SEVEN TIMES he shall dip his Finger in the BLOOD of a BULLOCK.

SEVEN TIMES I will smite you for your SINS.

In DEUTERONOMY, SEVEN PEOPLE possessed the LAND of PROMISE.

There, is also read, a SEVENTH YEAR of REMISSION.

And (there are) SEVEN CANDLES set up on the SOUTH SIDE of the CANDLESTICKS.

And in NUMBERS it is read, that the SONS of ISRAEL offered up SEVEN EWE LAMBS WITHOUT SPOT and that SEVEN DAYS, they did eat UNLEAVENED BREAD.

And that, SIN was expiated with SEVEN LAMBS and a GOAT.

And that, the SEVENTH DAY was Celebrated and HOLY.

And that, the First Day of the SEVENTH MONTH was observed and kept HOLY.

And the SEVENTH MONTH of the FEAST of TABERNACLES (was observed and kept HOLY).

And SEVEN CALVES were offered on the SEVENTH DAY.

And BAALAM erected SEVEN ALTARS.

SEVEN DAYS he touched a DEAD CARCASS was UNCLEAN.

And in JOSHUA, SEVEN PRIESTS carried the ARK of the COVENANT before the HOST.

And SEVEN DAYS they went round the CITIES.

And SEVEN TRUMPETS were carried by the SEVEN PRIESTS.

And on the SEVENTH DAY, the SEVEN PRIESTS sounded the Trumpets.

And in the BOOK of JUDGES, ABESSA reigned in ISRAEL SEVEN YEARS.

SAMPSON kept his NUPTIALS, SEVEN DAYS and the SEVENTH DAY he put forth a RIDDLE to his Wife.

He was BOUND with SEVEN GREEN WITHES (WILLOWS).

SEVEN LOCKS of his Head were shaved off.

SEVEN YEARS were the CHILDREN of ISRAEL oppressed by the KING of MADEN.

And in the BOOKS of the KINGS, ELIAS prayed SEVEN TIMES and at the SEVENTH TIME (he) beheld a Little Cloud.

SEVEN DAYS the CHILDREN of ISRAEL pitched over against the SYRIANS and in the SEVENTH DAY of the BATTLE, were joined.

SEVEN YEARS, FAMINE was threatened to DAVID for the People's Murmuring.

And SEVEN TIMES the Child sneezed that was raised by ELISHA.

And SEVEN MEN were Crucified together, in the DAYS of the FIRST HARVEST.

NAAMAN was made Clean with SEVEN WASHINGS, by ELISHA.

The SEVENTH MONTH, GOLIATH was Slain.

And in ESTHER, we read that the KING of PERSIA had SEVEN EUNUCHS.

And in TOBIAS, SEVEN MEN were coupled with SARAH, the DAUGHTER of RAGUEL.

And, in DANIEL, NEBUCHADNEZZAR'S FURNACE was heated SEVEN TIMES hotter than it was used to be.

And SEVEN LIONS were in the Den and the SEVENTH DAY came NEBUCHADNEZZAR.

In the BOOK of JOB< there is mention of SEVEN SONS of JOB.

And SEVEN DAYS and NIGHTS JOB'S Friends sat with him on the EARTH.

And in the same place, "In SEVEN TROUBLES, no EVIL shall come near you."

In EZRA, we read of ARTAXERXES' SEVEN COUNSELLORS and in the same place, the Trumpet sounded.

The SEVENTH MONTH of the FEAST of TABERNACLES was, in EZRA'S TIME, while the CHILDREN of ISRAEL were in the CITIES.

And on the First Day of the SEVENTH MONTH, ESDRAS read the LAW to the PEOPLE.

And in the PSALMS, DAVID praised the LORD SEVEN TIMES a DAY.

SILVER is tried SEVEN TIMES.

And he renders to his neighbors SEVEN-FOLD into their Bosoms.

And Solomon says, that WISDOM has hew herself SEVEN PILLARS.

(There are) SEVEN MEN that can render a REASON.

(There are) SEVEN ABOMINATIONS which the LORD abhors.

(There are) SEVEN ABOMINATIONS in the HEART of an ENEMY.

(There are) SEVEN OVERSEERS.

(There are) SEVEN EYES BEHOLDING.

ISAIAH numbers up SEVEN GIFTS of the HOLY GHOST.

And, (it is there a metaphor and meaningful prophesy that) SEVEN WOMEN shall take hold on a MAN.

And in JEREMIAH, (it is written that) if she that has borne SEVEN, languishes, she has given up the GHOST.

In EZEKIEL, the PROPHET continued sad for SEVEN DAYS.

In ZACHARIAH, (there are) SEVEN LAMPS and SEVEN PIPES to those SEVEN LAMPS.

And (there are) SEVEN EYES running to and from through the WHOLE EARTH.

And (there are) SEVEN EYES on ONE STONE.

And the FAST of the SEVENTH DAY is turned into JOY.

And in MICAH, SEVEN SHEPHERDS are raised against the ASSYRIANS.

Also, in the GOSPEL, we read of SEVEN BLESSINGS and SEVEN VIRTUES, to which SEVEN VICES are opposed.

(There are) SEVEN PETITIONS of the LORD'S PRAYER.

(There are) SEVEN WORDS of CHRIST upon the CROSS.

(There are) SEVEN WORDS of the BLESSED VIRGIN MARY.

(There are) SEVEN LOAVES distributed by the LORD.

(There are) SEVEN BASKETS of FRAGMENTS.

(There are) SEVEN BROTHERS having ONE WIFE.

(There are) SEVEN DISCIPLES of the LORD who were FISHERS.

(There are) SEVEN WATER POTS in CANA of GALILEE.

(There are) SEVEN WOES which the LORD threatens to HYPOCRITES.

(There are) SEVEN DEVILS cast out of the UNCLEAN WOMAN and SEVEN MORE WICKED DEVILS taken in after that which was CAST OUT.

Also, SEVEN YEARS, CHRIST was FLED into EGYPT.

And, the SEVENTH HOUR (in the Gospel), the FEVER LEFT the GOVERNOR'S SON.

And in the CANONICAL EPISTLES, JAMES describes SEVEN DEGREES of WISDOM.

And PETER (describes) SEVEN DEGREES of VIRTUES.

And in the ACTS, we reckon SEVEN DEACONS.

And (there are) SEVEN DISCIPLES CHOSEN by the APOSTLES.

Also, in the REVELATION, there are many MYSTERIES relating to this NUMBER, for there we read of ~

SEVEN CANDLESTICKS,
SEVEN STARS,
SEVEN CROWNS,
SEVEN CHURCHES,
SEVEN SPIRITS BEFORE THE THRONE,
SEVEN RIVERS OF EGYPT,
SEVEN SEALS, SEVEN MARKS,
SEVEN HORNS, SEVEN EYES,
SEVEN SPIRITS of GOD,
SEVEN ANGELS with SEVEN TRUMPETS,
SEVEN HORNS of the DRAGON,
SEVEN HEADS of the DRAGON,
    which had SEVEN DIADEMS,
also SEVEN PLAGUES and SEVEN VIALS,
    which were given to every one of the SEVEN ANGELS,
SEVEN HEADS of the SCARLET BEAST,
SEVEN MOUNTAINS and
SEVEN KINGS sitting upon them
and SEVEN THUNDERS uttered their VOICES.

Moreover, this NUMBER has much POWER, as in NATURAL, so in SACRED CEREMONIAL and also in other things.

Therefore, the SEVEN DAYS are related hither (in REVELATION).

Also the SEVEN PLANETS, the SEVEN STARS called PLEIADES, the SEVEN AGES of the WORLD, the SEVEN CHANGES of MAN, the SEVEN LIBERAL ARTS, and as many MECHANIC and so many FORBIDDEN (are related in REVELATION).

(There are) SEVEN COLORS, SEVEN METALS, SEVEN HOLES in the HEAD of a MAN, SEVEN PAIRS of NERVES, SEVEN MOUNTAINS in the CITY of ROME, SEVEN ROMAN KINGS, SEVEN CIVIL WARS, SEVEN WISEMEN in the Time of JEREMIAH, SEVEN WISEMEN of GREECE.

Also, ROME did burn SEVEN DAYS by NERO.
By SEVEN KINGS were Slain 10,000 MARTYRS.

There were SEVEN SLEEPERS and SEVEN PRINCIPAL CHURCHES of ROME.

DIAGRAM OF THE SCALE OF THE NUMBER SEVEN

1. In the Original World ~

2. Ararita ~

3. Adhadia ~

6. Asser Eheie ~

7 & 8. "I Shall Be" ~

9. The Name of God with Seven Letters ~

---

1. In the Intelligible World ~

2. Zaphiel ~

3. Zadkiel ~

4. Camael ~

5. Raphael ~

6. Haniel ~

7. Michael ~

8. Gabriel ~

9. Seven Angels which Stand in the Presence of God ~

Diagram of the Scale of the NUMBER SEVEN (continued)

In the Celestial World ~

Saturn ~
Jupiter ~
Mars ~
The Sun ~
Venus ~
Mercury ~
The Moon ~

Seven Planets ~

---

1. In the Elementary World ~

2. The Lapwing, the Cuttle-Fish, the Mole, Lead, the Onyx ~
3. The Eagle, the Dolphin, the Hart, Tin, the Sapphire ~
4. The Vulture, the Pike, the Wolf, Iron, the Diamond ~
5. The Swan, the Sea Calf, the Lion, Gold, the Carbuncle ~
6. The Dove, Thymallus, the Goat, Copper, the Emerald ~
7. The Stork, the Mullet, the Ape, Quick Silver, the Agates ~
8. The Owl, the Sea Cat, Cat, Silver, Chrystal ~

9. Seven Birds of the Planets,
   Seven Fish of the Planets,
   Seven Animals of the Planets,
   Seven Metals of the Planets,
   Seven Stones of the Planets ~

Diagram of the Scale of the NUMBER SEVEN (continued)

1. In the Lesser World ~

2. The Right Foot,
   The Right Ear ~

3. The Head,
   The Left Ear ~

4. The Right Hand,
   The Right Nostril ~

5. The Heart,
   The Right Eye ~

6. The Privy Members,
   The Left Nostril ~

7. The Left Hand,
   The Mouth ~

8. The Left Foot,
   The Left Eye ~

9. Seven Integral Members Distributed to the Planets.
   Seven Holes of the Head Distributed to the Planets. ~

Diagram of the Scale of the NUMBER SEVEN (continued)

1. In the Infernal World ~

2. Hell ~

3. The Gates of Death ~

4. The Shadow of Death ~

5. The Pit of Destruction ~

6. The Clay of Death ~

7. Perdition ~

8. The Depth of the Earth ~

9. Seven Habitations of Infernals, which Rabbi Joseph of Castilia, the Cabalist, describes in the Garden of Nuts. ~

Chapter 23
Of the NUMBER EIGHT and Its Scale

The PYTHAGOREANS call EIGHT the NUMBER OF JUSTICE and FULLNESS.

(This was among them thus) first, because it is First of all (those others) divided into NUMBERS EQUALLY EVEN, viz. into FOUR.

And that DIVISION is, by the same Reason, made into TWO times TWO, viz. TWO times TWO, TWO times.

And by Reason of this EQUALITY of DIVISION, it took itself the name of JUSTICE.

But the other received the name of FULLNESS, by Reason of the Contexture of the CORPOREAL SOLIDITY, since the First makes a SOLID BODY.

Hence that Custom of ORPHEUS swearing by the EIGHT DEITIES, if at any time he would beseech DIVINE JUSTICE, whose names are these ~ FIRE, WATER, EARTH, the HEAVEN, MOON, SUN, PHANES and the NIGHT.

There are only EIGHT VISIBLE SPHERES of the HEAVENS.

Also, by it, the PROPERTY of the CORPOREAL NATURE is signified, which ORPHEUS comprehends in EIGHT of his SEA Songs.

This is also called the COVENANT or CIRCUMCISION, which was commanded to be done by the JEWS (on) the EIGHTH DAY.

There were also, in the OLD LAW, EIGHT COMMANDMENTS of the PRIEST, viz. a BREASTPLATE, a COAT, a GIRDLE, a MITRE, a ROBE, an EPHOD, a GIRDLE of the EPHOD, and a GOLD PLATE.

Hither belongs the NUMBER to ETERNITY, and the END of the WORLD, because it follows the NUMBER SEVEN, which is the MYSTERY of TIME. Hence, also, the NUMBER OF BLESSEDNESS, as you may see in MATTHEW.

The NUMBER EIGHT is also called the NUMBER OF SAFETY and CONSERVATION, for there were so many SOULS of the SONS of JESSE, from which DAVID was the EIGHTH SON.

DIAGRAM OF THE SCALE OF THE NUMBER EIGHT

---

1. The Name of God with Eight Letters ~

2. Eloa Vaddath Jehova Vedaath ~

    (Elohim, Jehovah, Gabriel, Michael, Israel, Jacob, David, Solomon)

3. through 9. Eloa Vaddath Jehova Vedaath ~

10. In the Original World ~

Diagram of the Scale of the NUMBER EIGHT (continued)

1. Eight Rewards of the Blessed ~ (Empirical Worlds) ~

2. Inheritance ~
3. Incorruption ~
4. Power ~
5. Victory ~
6. The Vision of God ~
7. Grace ~
8. A Kingdom ~
9. Joy ~

10. In the Intelligible World ~

1. Eight Visible Heavens ~ (Celestial Heavens) ~

2. The Starry Heaven ~
3. The Heaven of Saturn ~
4. The Heaven of Jupiter ~
5. The Heaven of Mars ~
6. The Heaven of the Sun ~
7. The Heaven of Venus ~
8. The Heaven of Mercury ~
9. The Heaven of the Moon ~

10. In the Celestial World ~

Diagram of the Scale of the NUMBER EIGHT (continued)

1. Eight Particular Qualities ~ (Composite and Elemental) ~

2. The Dryness of the Earth ~
3. The Coldness of Water ~
4. The Moisture of Air ~
5. The Heat of Fire ~
6. The Heat of Air ~
7. The Moisture of Water ~
8. The Dryness of Fire ~
9. The Coldness of Earth ~

10. In the Elementary World ~

---

1. Eight Kinds of Blessed Men ~ (Rational) ~

2. The Peace Makers ~
3. They that Hunger and Thirst after Righteousness; the Passionate ~
4. The Meek; and Gentile ~
5. They which are Persecuted for Righteousness' Sake; the Martyrs ~
6. The Pure in Heart; the Pure and Divine ~
7. The Merciful; and Meek ~
8. The Poor in Spirit; the Humble and Contrite ~
9. The Mourners; the Faithful and True ~

10. In the Lesser World ~

Diagram of the Scale of the NUMBER EIGHT (continued)

1. Eight Punishments of the Damned ~
   (Indicted, Interdicted, and Interred) ~

2. Prison ~
3. Death ~
4. Judgment ~
5. The Wrath of God ~
6. Darkness ~
7. Indignation ~
8. Tribulation ~
9. Anguish ~

10. In the Infernal World ~

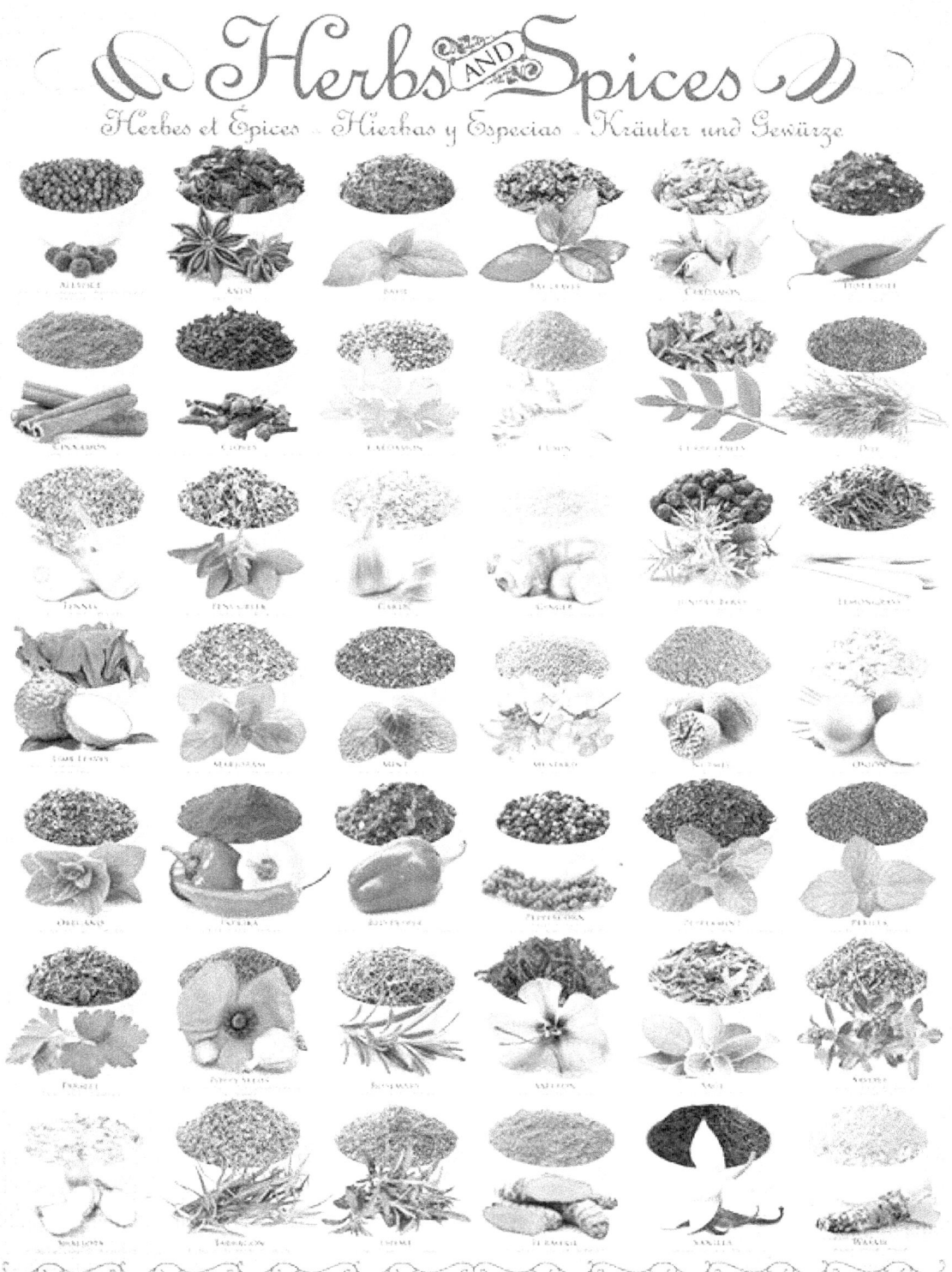

Chapter 24
Of the NUMBER NINE and Its Scale

There are NINE ORDERS of BLESSED ANGELS, viz. SERAPHIM, CHERUBIM, THRONES, DOMINATIONS, POWERS, VIRTUES, PRINCIPALITIES, ARCHANGELS, and ANGELS, which EZEKIEL figures out by the NINE STONES, which are SAPPHIRE, EMERALD, CARBUNCLE, BERYL, ONYX, CHRYSOLITE, JASPER, TOPAZ and SARDIS.

This NUMBER has also a Great and OCCULT MYSTERY of the CROSS, for the NINTH HOUR, OUR LORD JESUS CHRIST breathed out HIS SPIRIT.

The ASTROLOGERS also take notice of the NUMBER NINE in the AGES of MEN, no otherwise than they do of SEVEN, which they call CLIMACTERIAL YEARS, which are eminent for some remarkable change.

Yet sometimes, it signifies imperfectability and incompletion, because it does not attain to the PERFECTION of the NUMBER TEN, but less by ONE, without which it is DEFICIENT, as Austin (AUGUSTINE) interprets it of the TEN LEPERS.

Neither is the LONGITUDE of NINE CUBITS of OG, KING OF BASAN, who is a type of the DEVIL without a MYSTERY.

DIAGRAM OF THE SCALE OF THE NUMBER NINE

1. The Name of God with Nine Letters ~ (the Creator) ~

2. – 4. Jehovah Sabboath ~

5. – 7. Jehovah Zidkenu ~

8. – 10. Elohim Giber ~

11. In the Original World ~

---

1. Nine Choirs of Angels, Nine Angels Ruling the Heavens ~

2. Serpahim, Merattron ~
3. Cherubim, Ophaniel ~
4. Thrones, Zaphkiel ~

5. Dominations, Zadkiel ~
6. Powers, Camael ~
7. Virtues, Raphael ~

8. Principalities, Haniel ~
9. Archangels, Michael ~
10. Angels, Gabriel ~

11. In the Intelligible World ~ (in the Transcending Universal) ~

Diagram of the Scale of the NUMBER NINE (continued)

---

1. Nine Moveable Spheres ~ (Astrophysical and Phenomenal)

2. The Primum Mobile ~ (The Primal Labyrinth & The Firmament)
3. The Starry Heaven ~ (The Seven Higher Heavens)
4. The Sphere of Saturn ~ (The Lower Universe)

5. The Sphere of Jupiter ~
6. The Sphere of Mars ~
7. The Sphere of the Sun ~ (The Cosmogenic Universe under Heaven)

8. The Sphere of Venus ~
9. The Sphere of Mercury ~
10. The Sphere of the Moon ~ (The Physical Cosmogony; the Magical)

11. In the Celestial World ~

---

1. Nine Stones Representing the Nine Choirs of Angels ~
   (the Mithra of the Elemental Spirits) ~

2. Sapphire ~
3. Emerald ~
4. Carbuncle ~
5. Beryl ~
6. Onyx ~
7. Chrysolite ~
8. Jasper ~
9. Topaz ~
10. Sardis ~

11. In the Elementary World ~

Diagram of the Scale of the NUMBER NINE (continued)

1. Nine Senses ~ Inward and Outward Together ~
   (Mental and Physical; making Psychical) ~

2. Memory ~ (Sense and Image) ~
3. Cogitative ~ (Cognitive and Incognitive; Active and Passive) ~
4. Imaginative ~ (Faith, Mythology mingled) ~

5. Common Sense ~ (Rationalism) ~
6. Hearing ~
7. Seeing ~

8. Smelling ~
9. Tasting ~
10. Touching ~

11. In the Lesser World ~
    (In the Dominions of the Natural Intelligences) ~

Diagram of the Scale of the NUMBER NINE (continued)

1. Nine Orders of Devils ~
   (Nine Deadly Sins Personified in the Senses) ~

2. False Spirits ~
3. Spirits of Lying ~
4. Vessels of Iniquity ~ (Death) ~

5. Avengers of Wickedness ~
6. Jugglers ~
7. Airy Powers ~ (Hell) ~

8. Furies Sowing Mischief ~
9. Sifters or Triers ~
10. Tempters or Ensnarers ~ (Torment) ~

11. In the Infernal World ~
    (In the Meddling of Fire and Nature) ~

parsley sage rosemary thyme

Chapter 25
Of the NUMBER TEN and Its Scale

The NUMBER TEN is called EVERY NUMBER, or a UNIVERSAL NUMBER, Complete, signifying the Full Course of Life.

For beyond that we cannot Number but by Replication; and it either implies all Numbers within itself or explains them by itself and its own, by multiplying them.

Wherefore, it is accounted to be of Manifold Religion and Power and is applied to the PURGING of SOULS. Hence, the ANCIENTS called Ceremonies, DENARI (or 10's), because they were to be expiated and to offer Sacrifices and were to abstain from some certain things for TEN DAYS.

(DENARI in the most ancient days was also a term to refer to a complete set of Virtues.)

There are TEN SANGUINE PARTS of MAN ~ the menstrues (the female body gender; so DNA), the sperm (the male body gender; also DNA), the plasonatic spirit (by this is probably meant, the both platonic and passionate alternatively, so the emotional, however, the humors might be the inherent emotions and not those learned), the mass (the body in which the organic body is stored and carried and given place), the humors (have covered in spirit), the organic body (the organisms which support the reason, the mind, the emotion and the senses), the vegetable part (generally visceral), the sensitive part (this means the rational and automatic senses and organisms in the sensory range and not the emotions), reason (the complex mind) and the mind (the basic functions between emotions and senses).

(But these are only Nine Parts considering the first two are dividing mankind by gender; in which case, it could be that within the composite of man as different from beasts, the spine might be included and also the differentiable limbs; so that, there could be specificity and differentiation as female and male; this is only a suggestion; and if that is not good, then I would submit, the soul, which I know to be in the old lexicon of the 10 and also the spine again, because it establishes man separate of beasts, without his other natures, so in body as different and not just in mind.)

There are also, TEN SIMPLE INTEGRAL PARTS constituting MAN ~ the bone, the cartilage, the nerve, the fiber, the ligament, the artery, the vein, the membrane, the flesh and the skin.

There are also, TEN INTRINSIC COMPOSITE PARTS of MAN ~ the spirit, the brain, the lungs, the heart, the liver, the gall, the spleen, the kidneys, the testicles and the matrix.

(So by this I am guessing we could consider the sexual organs as good as the testes and the matrix being the entire group of organs in the visceral group which are besides the mass and flesh of the body, as which are also in the Integral Parts, such as the bone and the muscle and how it is organized as a defined and differential matrix. It is upsetting to me that he does not separate Man from the Beasts on a biological level. Matrix is another ancient word for the composite differentiation of man from beast, since in magic, changelings – men who were beasts and beasts who were men – were the primary reason for all morals against a life of plagues.)

There are TEN CURTAINS in the TEMPLE and TEN STRINGS in the PSALTERY and TEN MUSICAL INSTRUMENTS with which the PSALMS were sung, the names where were ~ neza, on which their odes were sung; nablum, the same as organs upon which they sang and played; mizmor, on which the Psalms were sung; sir, on which the Canticles were sung; tehila, on which orations were made; beracha, on which benedictions were given; halel upon which praises were accomplished; hodaia upon which thanks were made; asre upon which the felicity of any one could be recounted; hallelujah, upon which the praises of God only were given and also His contemplations.

There were also TEN SINGERS of PSALMS, viz. Adam, Abraham, Melchizedeck, Moses, Asaph, David, Solomon and the Three Sons of Chora; (the last of which we interpret as the Three Sons of Noah, who were Shem, Ham and Japheth).

There are also, TEN COMMANDMENTS.

And the TENTH DAY after the ASCENSION of CHRIST, the HOLY GHOST came down.

Lastly, this is the NUMBER, in which JACOB, wrestling with the ANGEL all night, overcame and, at the rising of the SUN, was blessed and called by the name of ISRAEL.

In this NUMBER TEN, JOSHUA overcame 31 KINGS and DAVID overcame GOLIATH and the PHILISTINES. And DANIEL escape the danger of the LIONS.

This NUMBER TEN is also circular, as UNITY. Because, being heaped together, it returns into a UNITY, from whence it had its Beginning. And it is the END and PERFECTION of all NUMBERS and the BEGINNING of TENS.

As the NUMBER TEN flows back into a UNITY from whence it proceeded, so everything that is flowing, is returned back to that from which it had the BEGINNING of its FLUX.

So water returns to the SEA from whence it had its beginning. The BODY returns to the EARTH from whence it was taken. TIME returns into ETERNITY from whence it flowed. The SPIRIT shall return to GOD who gave it. And lastly, every CREATURE returns to NOTHING from whence it was created.

Neither is it supported but by the WORD of GOD, in whom ALL THINGS are HIDDEN and ALL THINGS with the NUMBER TEN and by the NUMBER TEN, make a ROUND; as Proclus says, taking their BEGINNING from GOD and ENDING in HIM.

GOD, therefore (that FIRST UNITY, or ONE THING), before He communicated Himself to Inferiors, diffused Himself First into the FIRST of NUMBERS, viz. the NUMBER THREE, then into the NUMBER TEN, as into TEN IDEAS and MEASURES of MAKING ALL NUMBERS and ALL THINGS, which the Hebrews call TEN ATTRIBUTES and account TEN DIVINE NAMES; from which Cause there cannot be a further NUMBER.

Hence, ALL TENS have some DIVINE THING in them and in the LAW are required as HIS OWN, together with the FIRSTFRUITS, as the ORIGINAL of ALL THINGS and BEGINNING of NUMBERS and EVERY TENTH is as the END given to HIM, who is the BEGINNING and END of ALL THINGS.

DIAGRAM OF THE SCALE OF THE NUMBER TEN

1. In the Original ~

2. – 5. The Name of Jehovah of Ten Letters Collected ~

6. – 7. The Name of Jehovah of Ten Letters ~

8. – 9. Extended ~

10. – 12. The Name Elohim Sabaoth ~

13. The Name of God with Ten Letters ~

---

2. Eheie, Kether ~
3. Jod Jehovah, Hochmah ~
4. Jehovah Elohim, Binah ~
5. El, Hesed ~
6. Elohim Gibor, Geburah ~
7. Eloha, Tiphereth ~
8. Jehovah Saboath, Nezah ~
9. Elohim Saboath, Hod ~
10. Sadai, Jesod ~
11. Adonai melech, Malchuth ~
12. Ten Names of God.
    Ten Sephiroth.

Diagram of the Scale of the NUMBER TEN (continued)

1. In the Intelligible World ~
2. Seraphim, Hajothhakados, Merattron ~
3. Cherubim, Orphanim, Jophiel ~
4. Thrones, Aralim, Zaphkiel ~
5. Dominations, Hasmallim, Zadkiel ~
6. Powers, Seraphim, Camael ~
7. Virtues, Malachim, Raphael ~
8. Principalities, Elohim, Haniel ~
9. Archangels, Ben Elohim, Michael ~
10. Angels, Cherubim, Gabriel ~
11. Blessed Souls, Issim, the Soul of Messiah ~
12. Ten Orders of the Blessed according to Dionysius ~
    Ten Orders of the Blessed according to the Traditions of Men ~
    Ten Angels Ruling ~

1. In the Celestial World ~
2. Reschith Hagallalim, the Primum Mobile ~
3. Masloth, the Sphere of Zodiac ~
4. Sabbathi, the Sphere of Saturn ~
5. Zedeck, the Sphere of Jupiter ~
6. Madim, the Sphere of Mars ~
7. Schemes, the Sphere of the Sun ~
8. Noga, the Sphere of Venus ~
9. Cochab, the Sphere of Mercury ~
10. Levanah, the Sphere of the Moon ~
11. Holom Jesodoth, the Sphere of the Elements ~
12. Ten Spheres of the World ~

Diagram of the Scale of the NUMBER TEN (continued)

1. In the Elementary World ~
2. A Dove ~
3. A Lizard ~
4. A Dragon ~
5. An Eagle ~
6. A Horse ~
7. Lion ~
8. Man ~
9. The Fox ~
10. Bull ~
11. Lamb ~
12. Ten Animals Consecrated to the Gods ~

1. In the Lesser World ~
2. Spirit ~
3. Brain ~
4. Spleen ~
5. Liver ~
6. Gall ~
7. Heart ~
8. Kidneys ~
9. Lungs ~
10. Genitals ~
11. Matrix ~
12. Ten Parts Intrinsic of Man ~

Diagram of the Scale of the NUMBER TEN (continued)

1. In the Infernal World ~
2. False Gods ~
3. Lying Spirits ~
4. Vessels of Iniquity ~
5. Revengers of Wickedness ~
6. Jugglers ~
7. Airy Powers ~
8. Furies, the Seminaries of Evil ~
9. Sifters, or Triers ~
10. Tempters or Ensnarers ~
11. Wicked Souls Bearing Rule ~
12. Ten Orders of the Damned ~

Chapter 26
Of the NUMBERS ELEVEN and TWELVE, with the Cabalistic Scale

The NUMBER ELEVEN, as it exceeds NUMBER TEN, which is the NUMBER of the COMMANDMENTS, so it falls short of the NUMBER TWELVE, which is of GRACE and PERFECTION; therefore, it is called the NUMBER of SINS, and the PENITENT.

Now the NUMBER TWELVE is DIVINE and that NUMBER whereby the CELESTIALS are measured. It is also, the NUMBER of SIGNS in the ZODIAC, over which there are TWELVE ANGELS as CHIEF, supported by the IRRIGATION of the GREAT NAME of GOD.

In TWELVE YEARS, also, JUPITER perfects His course and the MOON daily runs through TWELVE DEGREES.

There are, TWELVE CHIEF JOINTS in the BODY of MAN, viz. in hands, elbows, shoulders, thighs, knees and vertebrae of the feet.

There is, also, a GREAT POWER of the NUMBER TWELVE in DIVINE MYSTERIES.

GOD chose TWELVE FAMILIES of ISRAEL and set over them TWELVE PRINCES, so many STONES were placed in the Midst of JORDAN.

And GOD commanded that so many (in NUMBER of TWELVE STONES) should be set on the BREAST of the PRIEST.

TWELVE LIONS did bear the BRAZEN SEA that was made by SOLOMON.

There were so many (of the NUMBER of TWELVE) FOUNTAINS in HELIM.

And so many APOSTLES of CHRIST set over the TWELVE TRIBES and TWELVE THOUSAND PEOPLE were set apart and chosen.

DIAGRAM OF THE SCALE
OF THE NUMBERS ELEVEN AND TWELVE

---

1. The Names of God with Twelves Letters ~
4. Holy ~
5. Blessed ~
6. He ~
10. Father, Son, Holy Ghost ~

Diagram of the Scale of the NUMBERS ELEVEN and TWELVE (continued)

1. The Great Name returned back into Twelve Banners ~
2. Jehovah ~
3. Yahweh ~
4. Jovan, Joba ~
5. My God ~
6. Hoya, Goya ~
7. Antiquity, Ages ~
8. Hallelujah, Majesty ~
9. Lord ~
10. Master, Sage ~
11. Esse, Jesse ~
12. Hebe ~

1. Twelve Orders of Blessed Spirits ~
2. Seraphim ~
3. Cherubim ~
4. Thrones ~
5. Dominations ~
6. Powers ~
7. Virtues ~
8. Principalities ~
9. Archangels ~
10. Angels ~
11. Innocents ~
12. Martyrs ~

Diagram of the Scale of the NUMBERS ELEVEN and TWELVE (continued)

1. Twelve Angels Ruling over the Twelve Signs ~
2. Malchidial ~
3. Asmodel ~
4. Ambriel ~
5. Muriel ~
6. Verchiel ~
7. Hamaliel ~
8. Zuriel ~
9. Barbiel ~
10. Adnachiel ~
11. Hanael ~
12. Gabriel ~

1. Twelve Tribes ~
2. Dan ~
3. Ruben ~
4. Judah ~
5. Manasseh ~
6. Asher ~
7. Simeon ~
8. Issachar ~
9. Benjamin ~
10. Naphthalin ~
11. Gad ~
12. Zabulon ~

Diagram of the Scale of the NUMBERS ELEVEN and TWELVE (continued)

1. Twelve Prophets ~
2. Malachi ~
3. Haggai ~
4. Zachariah ~
5. Amos ~
6. Hosea ~
7. Micah ~
8. Jonah ~
9. Obadiah ~
10. Zephaniah ~
11. Nahum ~
12. Habukkuk ~

1. Twelve Apostles ~
2. Matthias ~
3. Thaddeus ~
4. Simon ~
5. John ~
6. Peter ~
7. Andrew ~
8. Bartholomew ~
9. Philip ~
10. James the Elder ~
11. Thomas ~
12. Matthew ~

Diagram of the Scale of the NUMBERS ELEVEN and TWELVE (continued)

1. Twelve Signs of the Zodiac ~
2. Aries ~
3. Taurus ~
4. Gemini ~
5. Cancer ~
6. Leo ~
7. Virgo ~
8. Libra ~
9. Scorpius ~
10. Sagittarius ~
11. Capricorn ~
12. Aquarius ~

1. Twelve Months ~
2. March ~
3. April ~
4. May ~
5. June ~
6. July ~
7. August ~
8. September ~
9. October ~
10. November ~
11. December ~
12. January ~

Diagram of the Scale of the NUMBERS ELEVEN and TWELVE (continued)

1. Twelve Plants ~
2. Sang ~
3. Upright Vervain ~
4. Bending Vervain ~
5. Comfrey ~
6. Lady's Seal ~
7. Calamint ~
8. Scorpion Grass ~
9. Mugwort ~
10. Pimpernel ~
11. Dock ~
12. Dragonwort ~

1. Twelves Stones ~
2. Sardonius ~
3. A Cornelian ~
4. Topaz ~
5. Calcedony ~
6. Jasper ~
7. Emerald ~
8. Beryl ~
9. Amethyst ~
10. Hyacith ~
11. Chrysophrasus ~
12. Chrystal ~

Diagram of the Scale of the NUMBERS ELEVEN and TWELVE (continued)

---

1. Twelve Principal Members ~
2. Head ~
3. Neck ~
4. Arms ~
5. Breast ~
6. Heart ~
7. Belly ~
8. Kidneys ~
9. Genitals ~
10. Hams ~
11. Knees ~
12. Legs ~

---

1. Twelve Degrees of the Damned and of the Devils ~
2. False Gods ~
3. Lying Spirits ~
4. Vessels of Iniquity ~
5. Revengers of Wickedness ~
6. Jugglers ~
7. Airy Powers ~
8. Furies, the Sowers of Evil ~
9. Sifters, or Triers ~
10. Tempters, or Ensnarers ~
11. Witches ~
12. Apostates ~

Chapter 27
Of the Notes of the HEBREWS and CHALDEANS
and Other Notes of MAGICIANS

    The HEBREW CHARACTERS have marks of NUMBERS attributed to them far more EXCELLENT than any other Language, since the Greatest Mysteries lie in the HEBREW LETTERS, as is handled concerning these in that Part of CABALA which we call NOTARIACON.

    Now the PRINCIPAL HEBREW LETTERS are in NUMBER, TWENTY-TWO, whereof, FIVE have various other Certain FIGURES in the END of a WORD, which therefore, they call the FIVE END LETTERS, which, being added to them aforesaid, make TWENTY-SEVEN. Which being then divided into THREE DEGREES, signifies UNITS, which are in the FIRST DEGREE ~ TENS, which are in the SECOND ~ and HUNDREDS, which are in the THIRD DEGREE.

    Now every one of them, if they are marked with a Great Character, signifies so many THOUSANDS, as (shown) here.

3000	2000	1000
ג	ב	א

The CLASSES of the HEBREW NUMBERS are these which follow.

1	2	3	4	5	6	7	8	9
א	ב	ג	ד	ה	ו	ז	ח	ט
10	20	30	40	50	60	70	80	90
י	כ	ל	מ	נ	ס	ע	פ	צ
100	200	300	400	500	600	700	800	900
ק	ר	ש	ת	ך	ם	ן	ף	ץ

Sometimes, the FINAL LETTERS are not used but we will write them thusly.

1000	900	800	700	600	500
א	קתת	תת	שת	רת	קת

And by those SIMPLE FIGURES and by the joining them together, they describe all other COMPOUND NUMBERS as ELEVEN, TWELVE, ONE HUNDRED and TEN, ONE HUNDRED and ELEVEN (110, 111), by adding to the NUMBER TEN those which are UNITS; and in the like manner to the rest, after their own manner.

Yet we describe the FIFTEENTH NUMBER not by TEN and FIVE, but by NINE and SIX, viz. וט . And that our of honor to the DIVINE NAME, יה, which signifies FIFTEEN, les that SACRED NAME should be abused to Profane Things.

Likewise the Egyptians, Ethiopians, Chaldeans and Arabians, have their Marks of NUMBERS, which serve for the making of MAGICAL CHARACTERS. But the CHALDEANS mark their NUMBERS with the LETTERS of their ALPHABET, after the manner of the Hebrews.

I found in a very ANCIENT BOOK of MAGIC, some very ELEGANT CHARACTERS, which I have figured in the following.

Now of these CHARACTERS, turned towards the LEFT HAND, are made TENS.

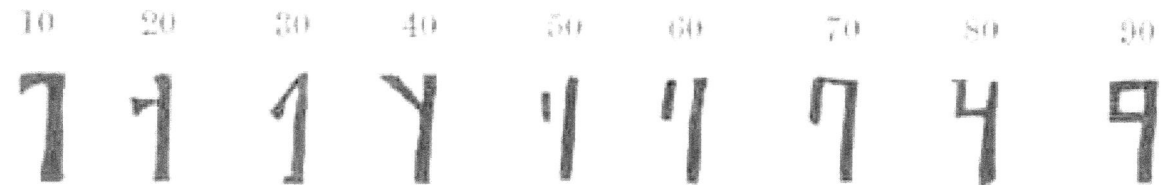

And those marks being downwards, to the RIGHT HAND, make HUNDREDS, to the LEFT HAND, THOUSANDS, as shown below.

And by the COMPOSITION and MIXTURE of these CHARACTERS, other COMPOUND NUMBERS are most elegantly made, as you may perceive by these few.

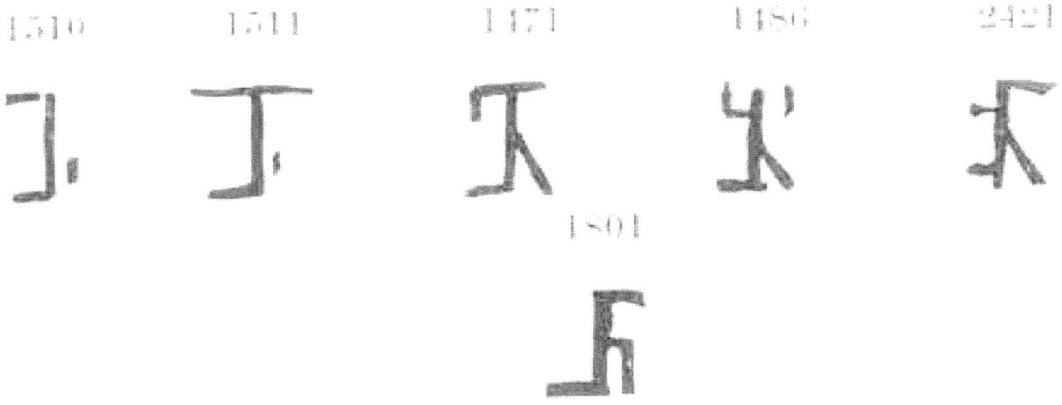

259

Plate 1

The Magic Tables, Seals & Characters of the Planets, their Intelligences & Spirits

The Table of Saturn in his Compass

4	9	2
3	5	7
8	1	6

The same Table in Hebrew

ד	ט	ב
ז	ה	ג
ח	א	ו

The Seal of Saturn   Of the Intelligence of ♄   Of the Spirit of ♄

The Table of Jupiter

4	14	15	1
9	7	6	12
5	11	10	8
16	2	3	13

In Hebrew

ד	יד	יה	א
ט	ז	ו	יב
ה	יא	י	ח
יו	ב	ג	יג

The Seal of Jupiter   Of the Intelligence of ♃   Of the Spirit of ♃

The Table of Mars

11	24	7	20	3
4	12	25	8	16
17	5	13	21	9
10	18	1	14	22
23	6	19	2	15

In Hebrew

יא	כד	ז	כ	ג
ד	יב	כה	ח	יו
יז	ה	יג	כא	ט
י	יח	א	יד	כב
כג	ו	יט	ב	יה

The Seal of Mars   Of his Intelligence   Of his Spirit

Designed by F. Barrett   Engraved by ———

Chapter 28
The MAGIC TABLES of the PLANETS ~
Their Form and Virtue ~
What are the Divine Names, Intelligences and Spirits over the Planets

There are CERTAIN MAGIC TABLES of NUMBERS distributed to the SEVEN PLANETS, which they call the SACRED TABLES of the PLANETS, because, being rightly formed, they are endued with the many GREAT VIRTUES of the HEAVENS, insomuch that they represent the DIVINE ORDER of the CELESTIAL NUMBERS, impressed upon them by the IDEAS of the DIVINE MIND, by means of the SOUL of the WORLD and the SWEET HARMONY of those CELESTIAL RAYS; signifying, according to proportion, SUPER-CELESTIAL INTELLIGENCES, which can no other way be expressed than by the marks of NUMBERS, LETTERS and CHARACTERS.

For MATERIAL NUMBERS and FIGURES can do nothing in the MYSTERIES of HIDDEN THINGS, but representatively by FORMAL NUMBERS and FIGURES, as they are governed and informed by INTELLIGENCES and DIVINE ENUMERATIONS, which unite the EXTREMES of the MATTER and SPIRIT to the WILL of the ELEVATED SOUL, receiving (through Great Affection by the Celestial Power of the Operator) a VIRTUE and POWER from GOD, applied through the SOUL of the UNIVERSE and the observation of CELESTIAL CONSTELLATIONS to a MATTER fit for a FORM, the Mediums being disposed by the Skill and Industry of the MAGICIAN.

But now we will hasten to explain each particular Table.

THE FIRST TABLE is assigned to the PLANET SATURN.

It consists of a SQUARE of THREE, containing the particular NUMBERS of NINE and in every LINE, THREE (numbers) every way. And through each DIAMETER making FIFTEEN, the whole SUM of NUMBERS, FORTY-FIVE (45).

Over this are set such DIVINE NAMES as fill up the NUMBERS with an INTELLIGENCE to what is GOOD and a SPIRIT to what is BAD. And out of the same NUMBERS are drawn the SEAL and CHARACTER of SATURN and of the SPIRITS thereof, such as is beneath ascribed to the TABLE.

Now this TABLE, being with a FORTUNATE SATURN, engraved on a PLATE of LEAD, helps Childbirth; and to make any Man safe or powerful; and to cause Success of Petitions with Princes and Powers. But if it be done, SATURN, being UNFORTUNATE, it hinders buildings, plantings and the like and casts a Man from honors and dignities and causes discord, quarrelling and disperses an army.

(Having never investigated this and being ashamed to rationally present some science which I cannot account for to any mystic values or metaphysical I might underestimate (this being a mathematical quiz and my attainment to that end not being excellent enough), my surmisal of this meaning, though I doubt the summation is correct thereafter – is in this interpretation – that the first square basis is given and then each side of the square has in it three numbers, each of them three in value, which together their sum is nine. So that the diameter of a circle inside of the square is this number fifteen by the meeting of the sides of a triangle protracted into it and measuring sides of five perhaps to all sides of the triangle to which the volume of the circle itself is measured the fifteen for the number completion. But it is probably arithmetic and not geometric and has no meaning to the actual principals which might allow it. And where there is the ability for the number forty-five is obviously by estimating that the diameter has a new mass volume ability within the triangle of the sphere which is three again. That allows in demesne for the fourth side of the square to establish itself as a moveable line throughout its diameter and its internal triangle when it has achieved a certain mass and volume. How this entire establishment becomes real or physical is of complete doubt to my imagination.)

THE SECOND TABLE is the TABLE of JUPITER.

It consists of a SQUARE drawn into itself. It contains SIXTEEN particular NUMBERS and in every LINE and DIAMETER, FOUR (numbers) making THIRTY-FOUR (34). The SUM of all (of it) is ONE HUNDRED and THIRTY-SIX (136).

There are over it, DIVINE NAMES, with an INTELLIGENCE to that which is GOOD and a SPIRIT to that which is BAD. And out of it is drawn the CHARACTER of JUPITER and the SPIRITS thereof.

If this TABLE is engraved on a PLATE of SILVER, with JUPITER being POWERFUL and RULING in the HEAVENS, it conduces to gain riches and favor, love, peace and concord and to appease enemies and to confirm honors, dignities and counsels. And it dissolves Enchantments if engraved on a CORAL.

(To have to try my hand at this a second time, I doubt I have a better course of logic than the first; in fact it is worse, severely. The basis which is the square then has the number four to each side and the total of it all is 136, so that possibly the primary square which is the basis is withdrawn through some power of the other numbers to exist within it and leave behind the sum of 120; into which sum may be divided to the sides the same leaving it to the sum of 30 to a side. Now I gave the sides each the number four instead of sixteen to assume there is some value in which each side may have a box of its own with four sides four, to make this subjective square the side of 16 numbers. It has no meaning I am sure, but I try to create a theory for the words. As for the number 34 it has no meaning otherwise since it is just the square itself taken in half and then added by the number of times again left which is 2 – perhaps this signifies the motion of the planet. Should it be known my mathematical habits are almost fictional, in that I enjoyed to read the story of mathematics and learned it in all my courses, thus. I studied the regular course of higher mathematics and was forced to decline to entertain the discipline as creative. The rules of it were too vagarious and multifarious to remember and to place to a more simple system of ordinary logic and it would not relate itself to story anymore.)

THE THIRD TABLES belongs to MARS.

The THIRD TABLET of MARS is made of a SQUARE of FIVE and it contains TWENTY-FIVE NUMBERS and of these, in every SIDE and DIAMETER, FIVE, which makes SIXTY-FIVE and the SUM of all is THREE HUNDRED and TWENTY-FIVE (325).

And there are over it, DIVINE NAMES with an INTELLIGENCE to GOOD and a SPIRIT to EVIL and out of it are drawn the CHARACTERS of MARS and of his SPIRITs.

These, with MARS FORTUNATE, being engraved on an IRON PLATE or (IRON, VULCAN) SWORD, makes a Man potent in war and judgment and petitions and terrible to his enemies and victorious over them. And if engraved upon the STONE CORREOLA, it stops blood and the menstrues. But if it is engraved with MARS being UNFORTUNATE, on a PLATE of RED BRASS, it prevents and hinders building; it casts down the powerful from dignities, honors and riches and causes discord and hatred among Men and Beasts. It drives away bees, pigeons and fish and hinders mills from working (or that is, it binds mills from working). It likewise renders hunters and fighters unfortunate and it causes barrenness in men and women alike and strikes a terror into our enemies and compels them to submit.

(Defiant to the end that I must use my magical metallic devise of instrumental magic and astrophysical principle, I must calculate this by a machine, not even knowing quite how to produce the proper results by the machine usage of the equations.

In which case without any machine of magic I begin by saying that the square of basis has the number fives to all four of its sides and so that the sum of the sides is also basic by arithmetic means and is twenty. The extra set of a five would be the vector resulting from the diameter of five which may be used spatially to move and expand the diameter by thirteen times its own intrinsic and basic measure. The sum, they say, of all the diameter in its basis of 20 is 300, so that, if we multiply the original number 65 of the mass sum of the diameter, then we get the number 260, which is void of 40 numbers of need. In which case, I would suggest rather to use a magical devise to finish my allusion to an equation, since I must confess I do not then know the method of proof thereof. But essentially, using the number 20 twice, we achieve the 40 void again. Now this is attributed to the mass sum within both the base and the diameter in order to come to their own conclusion. And obviously they have some method to re-endorse the diameter again to the mass sum, whereas, we endorsed it first and already to make all these conclusions of possibility. And the usage of the base 20 being used twice is again to show the division of the diameter into have and reapplied as many times as it is, which is twice. But though this is a rule, I have forgotten it as well by proof.)

THE FOURTH TABLE is of the SUN.

The FOURTH TABLET that is the SUN, is made of a SQUARE of SIX and contains THIRTY-SIX particular NUMBERS.

Of these THIRTY-SIX NUMBERS, SIX is in every SIDE and DIAMETER and they produce ONE HUNDRED and ELEVEN (111) and the SUM of all is SIX HUNDRED and SIXTY-SIX (666).

There are over it DIVINE NAMES, with an INTELLIGENCE to what is GOOD and a SPIRIT to what is EVIL. Out of it is drawn the CHARACTER of the SUN and of his SPIRITS.

This TABLE being engraved on a PLATE of PURE GOLD, being SOL FORTUNATE, renders him that wears it, renowned, amiable, acceptable, potent in all his works and equals him to a King; elevating his fortunes and enabling him to do whatever he will. But with SOL UNFORTUNATE, it makes one a tyrant, proud, ambitious, insatiable and finally that person to come to an ill-ending.

(The equities given for the calculations of the Sun in the mystery of the numbers, is symmetrical enough to prove may difficult points, which I cannot elucidate; those principally being the very ones of halves, quarters and thirds all in motion together in the geometric trigonometry which is a hyperbole of calculus and the foreign domain of chemical astrophysics as well. To which, if simpler mathematics and some first year algebra cannot answer (and unfortunately for the unlearned simpletons who have tried it, it does), there is no sure path of logic to believe. (However, it is hopeful for the former simpletons – I mean myself, to know that, investigating the matter, linear mathematics exceeds the triple threat of the virtuous by doctors.) So, although we should or at least could be ashamed of ourselves not to remember enough of both of our early disciplines and the basis of geometry to supersede the graduates of the keys, we do not have to be so much as to say equal or above them, because that would take a ridiculous amount of education, equal to a physicians' years.

Let us continue. With a story which makes an effort to elucidate how difficult the lessons are for the simple and what otherwise should be a simple exercise of correction for the advanced of learning.

So I say. Each side of the square basis will have the number six ascribed to its side as representative of its value of some kind of nature, as in all the other examples, where the number of value is ascribed to the side to become the integral factor of relevance and importance in the works of the mechanics of the equations attributed or attributable to the basis as written and inscribed by the number a pertinent value and one that is principally attributable to the force of magnetism in the mathematical and geometric schema, since we have the substance of the scheme, which is a planet and understand it not to be a static body in any case, but having astrophysical sensation which corresponds to measurable experiment resulting in data.

These sixes reveal the number 24. For the fact that in the other equations we were invited to increase the square to a pentangle force for the case of the vector to the hyperbole in order to create a diameter which may be spoken of in terms of equations relating to mass and volume, then I increase by automation of its method, the number to 30. This is now the magical basis of the Sun as it is handed over to us as a square. It's interior corresponds in effects to 30, from the raw basis of 24. We now need to understand how there is an equation which produces that 111 value from the number given at basis of itself and its ability for motion at 30. Needless to say, we will attempt to use the simple mathematics and additions where need to account for it and hope that this will respond to the basis. As for the total result of its deadly form of result in the very number of prophesy, 666, then we must only believe that the original conclusion of the number 36, which I have not yet spoken for, is employed to accepting that we began with the number 6 and that it had a quality in its mass and volume in total which had redemption value from the intermediate sum of 111 to reaction again in the whole and increasing the 111 value by itself at origin to its final causality. How this happens is the mystery of this literature.

First we will take the number 30 out of its vector class in order to establish that it will reduce or increase the sun to a volume of 3 times itself in either direction of size if it remains that the sum of 19 may be incorporated to another mass. This increases its volatility for motion according to the former principal of vector analysis. But we had binary analysis as well that was applied in the same rules and here we have no vector analysis so I have included the introduction of it.

Somewhere set in the value of the sun's square basis, there is no vector analysis implied, but the binary analysis is that only which proceeds, so the diameters are given to a rectangular section of the interior diameter and allows that the fifth vector is stationed in parallel symmetry with the 6th vector which should remain stationary, unless we are to suddenly allow that one vector implodes the sun by mass and volume and the other vector explodes the sun by mass and volume, according to the rule of 19, in which case, we may say we only have one vector but it occurs as two because of its alter-

magnetism; whereas, the magnetism of the other vectors were only explosive. But I have no mystery for the number 21 which renders itself, except that it is the number three, repeated six times in a sequence which occurs once only as three extra times. In which case, this is a very static and motioning body all at the same time and I would surmise that it might mean this planet (which is also called a star), is special because it has occurred in an altered universe in which such fantastic dimensional resolutions occur, and also seems to shine in our own. How may any stationed body continue to destroy and remake itself unless it were a constant fire burning stationed in some matrix of gravity which had its own separate dimension that was devastated upon entry into a foreign dimension? It could be the fictional gateway to the stars, in which case, this sun is not a star, but a fiery planet. Who could know except the aesthetician who built it?

    The explanation then for the six is given is its construction by placing a rectangle within the diameter and ascribing new 6's to each of those two parallel sides and not completely the top or bottom lines of the holographic rectangle, in order that the composition of the planet itself can be accounted for in its fiery exit and re-entry into the planetary range in which it finds itself with chemical affinities like itself. In which case, we can only achieve by this schema of addition again, to use our primitive wits, that the number 36 in 3 revolutions only once, as we have employed first the vector theory in anti-magnetism in the use of the number 30 with 21 left over as the combination of 3 used 6 times in only a single 3 sequence routine, perhaps as to explain the force of the internal fire as separate from the force of any other magnetism which could otherwise define the sun in the schema of the planetary domain, extrinsic of the stellar domain. But alas, we lose the struggle in that, the number 36 over three times its domain to disprove the anti-polarity of the diameter into a unified course again, is only 108 in number and we are left with a sad number 3 to account for failure.

So much for my story of the evil number 666. It's a disappointment by the number 3. But it may be a necessary value to add 3 to III to obtain 114 to signify that the magnet of the open rectangular parallel should be induced to becoming this evil, 666.)

THE FIFTH TABLE is of VENUS.

The FIFTH TABLET, which belongs to VENUS, consists of a SQUARE of SEVEN, drawn into itself, viz. of FORTY-NINE NUMBERS, whereof SEVEN on each SIDE AND DIAMETER make ONE HUNDRED and SEVENTY-FIVE and the SUM of all is ONE THOUSAND TWO HUNDRED and TWENTY-FIVE.

There are, likewise, over it, DIVINE NAMES, with an INTELLIGENCE to GOOD and a SPIRIT to EVIL. And there is drawn out of it, the CHARACTER of VENUS and her SPIRITS.

This TABLE, being engraved on a PLATE of SILVER, being such of the VENUS FORTUNATE, promotes concord, ends strife, procures the love of women, helps conception, is good against barrenness, gives ability for generation, dissolves enchantments, causes peace between man and woman and makes all kinds of animals fruitful and likewise, makes cattle fruitful. And being put into a dove or pigeons house, it causes an increase. It likewise drives away melancholy distempers and causes joyfulness. And this, being carried about by travelers, makes them fortunate. But if it is formed and engraved upon BRASS, this being of the VENUS UNFORTUNATE, it acts contrary to all that has been said.

(The mathematics become increasingly more complicated as the number of the tablets are increased by virtue of their names in order of their appearing to our subject. In the fifth tablet, we find the magnetic orders of the number 7 and whatever any magnetic order may have been before of the side values of the square basis, could remain mystery unless we are somewhere given the faulted advice of the chimerical answers for the motives of their values in an extra science; so now we are approached with the aesthetic value of 7.

To itself it is the number 28 at basis and by a vector singular it is 35 and by a vector anti-polar (or binary) it is 42; and however, we must find it as the number 49, in which case we need to invent a science for the terse or tertiary effect of the vectors, which is a stationed motion object of a moving triangle. This should make a good story sometime. But not now. So we will employ the vector as a motion triangle put at the point of the medium of the diameter and allow it to move us wherever it might be of best and greatest use in order to reach the desired effects of the remaining numbers, which are 175 and 1225 by dimensional applications of the vectors, first in single then in binary and finally in triple (terse-base), or more simply, as a movable triangle. And all those optional applications of vector analysis only being if need states that we must use them to achieve the correct geometric results to the answers.

Now I must use my magical metal devise for calculation.

So we find that 175 divided by 28 is 6 and one quarter.

And, 175 divided by 35 is 5, a nice equal rationale and half of perfection which should be helpful.

But alas, 175 divided by 42 is an unpleasant conclusion and I must report is as I have found it out by the magical metal, as 4.1666666. (We will leave that there, since it's unlikely that the binary system will be one that we will choose for Venus, since it is too complicated, using double halves and a singular additive and many double thirds; this is no proposition for any algebraist with a doctorate.

So we cannot include Venus in our plans with the Sun. But as for the singular vector account of the earlier planets, maybe.)

And in the case of which they give us this equation to follow, the result is still more horrible, which is the reported result of, 3.5714285; and this should have absolutely no meaning to anyone but a Venetian Doctor of Physics born thereof.

In which of all case we will allow ourselves the following schema of rules to equate realism with the possible conclusions and after the schema is quit, we will adapt a story to the total outcome.

The schema is this. We will use the simple vector rule to abscond the triangle at origin derived from lacking 3 sides to the square of basis.

So doing, our vector result by principle is the value of 5. Therefore, since the vector must be used to establish the force of the triangle into its place of action for the total results of 175 and 1125, we will employ the devices of the magical consequence of 3 and 5 with 7 to achieve any goal of its wit from a square named Venus.

And immediately we are unhappy ~ 175 divided by 3 is 58.333333. But we know we are happy with 5 being a result in division of 35 and finally, 175 divided by 7 is 25 which is also happy. We see the opportunity to put the PERFECT NUMBER into all of this work somehow. For if we apply the number of the principle in question, which is 7 to the same division of ultimate means by the result of a vector rule, giving the same the weight of 5 to establish the vector rule of 7 times 5 being 35, the vector rule achieving itself in the 5th dimension of the square in ascription. Then increasing the number to the end of its domain in 49 by passing itself to the 7 of demesne, then it will take us to a reduction of the vector value by TEN, into the number of resulting 25. We have had a TEN Scale dimensional change somehow by the values of vectors into origin by zero changes. And through the usage of binary objection and triangular occurrence which we say is more than probably something like tertiary acceptance.

Let us hope that maybe a word, otherwise, by a magical triangle which can hang on to a vector and stay put at least until the vector itself moves it, in order to occur into a tertiary dimension; since we have seen that the binary dimension is literally transcending.

Where and/or how should this magical number be used as concept or votive or force or theory. Should there be TEN angels at this point, I would be happy for us. But, by any means, I achieve the number 175 in passage of the motion which objects to binary dimension or hyperbole and accepts this tertiary dimension or hyperbole. But we need to transfer this tertiary hyperbole over into the number of 1225. Should be continue or give up? Let's continue a bit more to the extent of forgiving ourselves of the magical metal devise which we're obligated to using. And this we would have otherwise known immediately had we not needed the magical metal device, that, 1225 divided by 175 is seven. We have truly transcended.)

THE SIXTH TABLE is of MERCURY.

The SIXTH TABLET, which is that of MERCURY, results from a SQUARE of EIGHT drawn to itself, containing SIXTY-FOUR NUMBERS, whereof EIGHT on every SIDE by BOTH DIAMETERS, make TWO HUNDRED and SIXTY. And the SUM of all (of it) is TWO THOUSAND and EIGHTY.

Over the SIXTH TABLET are set DIVINE NAMES, with an INTELLIGENCE to GOOD and with a SPIRIT to BAD and from it is drawn a CHARACTER of MERCURY and the SPIRITS thereof.

If with MERCURY FORTUNATE you engrave it upon SILVER, TIN or YELLOW BRASS, or write it upon VIRGIN PARCHMENT, it renders the bearer thereof grateful, acceptable and fortunate to do whatever he pleases. It brings gain and prevents poverty and helps memory, understanding and divination and to the understanding of occult things as well, as by dreams. But with MERCURY UNFORTUNATE, it will perform everything in contrary to the good will of its fortune.

(It goes without saying or does it? We need a magical devise for this.

Every side obviously has eight drawn to it and each of those four sides of eight is duplicated in meter, or dimension by twice its own magnitude, so that each number eight in value, makes up one eighth part of a new number eight value which opposes the meter of the original square and becomes its binary self to create a number 64 value as is given. Whereas, the number 260 is by drawing the vector principle into the schema of what is established and finding that value by the additive component, which we will do shortly. And finally we have to account for the total values as being 2080, a pretty value if not also a useful number or vice versa.

So we need to be creative with our poor means; 2080 divided by 64 is 32 and a half and that is not rational so we try to dissuade its notion as being a correct application, except if and only if the half principle of the last rational number that is .5 and wants otherwise to be 1, throwing the entire calculation off, is useful to achieving the point of the eighth of a value point in the binary value of the 8's to the sides themselves.

There are other values to consider. If eight times four for the squares original basis weight is 32 and it is, then we have the basis to consider, along with its vector value which we figure by our theory as 40, as being a fifth side which does not occur and this would be a necessary value, that of 40 and not that of 32 because of our account into the number 64 by binary explanation.

So therefore, we also find that, 2080 is divisible by 40 to give us the value of 52, which is mystical in that it is also the number of the weeks of our terrestrial year on earth; however much, we also don't know to account if that is relevant data in correlation of ourselves to the planet of mercury. And we also see that the number 52 is only the number 12 in value from 64, which is usable to the extent of the tertiary over the basis or the triangular form over the square.

In which case, if we were to draw fixed triangles over the four corners of the mercurious square planet basis, we would establish those for our tertiary nature and remain in the square basis by the establishment of the excess of the diameters of 64 in the binary range which we however cannot seem to transcend. This is therefore good because that is all the construction we will need to close this discussion; all the data given has been accounted for.)

THE SEVENTH TABLE is of the MOON.
(It is the LAST TABLE.)

The SEVENTH TABLET which is for the MOON, consists of a SQUARE of NINE, having EIGHTY-ONE NUMBERS in every SIDE and (in) DIAMETER NINE, producing THREE HUNDRED and SIXTY-NINE and the SUM of all is THREE THOUSAND and TWENTY-ONE.

There are, over the SEVENTH TABLET, DIVINE NAMES, with an INTELLIGENCE to what is GOOD and a SPIRIT to EVIL. And from it are drawn the CHARACTERS of the MOON and the SPIRITS thereof.

This, the MOON FORTUNATE, engraved on SILVER, makes the bearer amiable, pleasant, cheerful and honored, removing all malice and ill-will. It causes security in a journey, increase of riches and health of body, drives away enemies and other evil things from what place soever you shall wish them to be expelled. But if it is the MOON UNFORTUNATE and it is engraved on a PLATE of LEAD, wherever it shall be buried, it makes that place unfortunate and the inhabitants thereabouts, as also ships, rivers, fountains and mills. And it makes every man unfortunate against whom it shall be directly done, making him fly his place or abode (and even his country), where it shall be buried. And it hinders physicians and orators and all men whatsoever in their office, against whom it shall be made.

Now how the SEALS and CHARACTERS of the PLANETS are drawn from these Tables, the Wise Searcher and he who shall understand the verifying of these Tables, shall easily find out.

Here follow the DIVINE NAMES corresponding with the NUMBERS of the PLANETS, with the NAMES of the INTELLIGENCES and DEMONS, or SPIRITS, subject to those NAMES.

It is to be understood that the INTELLIGENCES are the Presiding GOOD ANGELS that are set over the PLANETS, but that the SPIRITS or DEMONS, with their NAME SEALS or CHARACTERS, are never inscribed upon any TALISMAN, except to execute any EVIL effect and that they are subject to the INTELLIGENCES or GOOD SPIRITS. And again, when the SPIRITS and their CHARACTERS are used, it will be more conducive to the effect to add some DIVINE NAME appropriate to that effect which we desire.

(Were it not this to be accounted the final planet of report, it would be a most desperate hour for the aesthetician of couches and decorations; this will be seen to include myself someday.

Back to the magical calculations, we notice immediately that the vectoral number is ascribed in direct appropriation of the diameter in this account, whereas it is not before except in the line, so there are more things to be understood if we had some mastery over the line to the part of the diameter. But we don't. We have accounted lines and diameters all as the magic of linear vectors through the use of the tri-disciplines of trigonometry and calculus over the geometric basis; and as for us, not even that, we simply account the algebraic course in notation of these other methods as having rules accountable and return back to those methods of calculations with the additional values of arithmetic means in the values of the geometric ones.

Therefore we understand that we have a binary universe to account in the moon's composite form of these numbers because the diameter effects all the sides through the vector apparent and stated, which envelopes the domain with the substance of itself and is the definition of a binary universe, excepting that in heavier scale bodies, like the formers, the binary system does not actually decry itself through the given meters, it only implies that it might be present through the exercise of some rule. So whereas, the sun does transfer itself into an ephemeral dimension by the binary and the others have to obviate the binary dimensions to get at their own terse rules, then the moon is enveloped within the promise of this original ephemeras and remains hidden from it in the shadow of its possibility at the very core of its un-terse diametrical point which corresponds to all its square sides.

So our numbers are 9, 81, the notation of the diameter 9 again, 369 and 3021. Neither of the last two numbers are hopeful. So we hope that the double effect of the number 9 through the diameter in a stationed vector binary explanation will be pleasant to ending the journey to the moon finally and no more into the beyond. We are a long way away from the New Jerusalem at this point and this is the concern of many theologians regarding the occult sciences. Let us table the number itself which is the attribute itself and only note that it might be duplicated in error of its binary notice into the number 18 and again made terse into the number 27 because of its lacking a vector motion of indication besides itself, which, if it did, would resolve itself into the number 9 or 18 again, so there is no point in it; through the binary point, which is doubled. In which case, perhaps we could extend the numbers again to 18, 36 and 54. We will keep this progression in mind; for that they conclude again 9, 12 and 17 +3, which might be useful to make a change.

But we find a rational result using magic metals (that is in jest); that, 369 divided by 9 is 41, a rational result. I account that I have not recently studied the definition of rational numbers besides that they are not fractioned. Can this number 41 be helpful? Only so far as we are willing to be creative and see that the expansion by terse dimensions from using 9 by 3 as 27 and thrown into the binary so that it was 54 and reducing it again by the terse ability lacing of the stationed vector back to 17 +3, we see that 17 in the terse dimension is 51 and that is only the magic number ten higher than the result of the magnate of 41 into the 369 result 9 times. I will try no further to explain that I use 10 as the perfect resultant universe it wants to be, alleging that the moon and the sun are electromagnetic partners and share an anti-magnetic resolution in the binary, effected by the terse dimension, which may as well be the value of 10, having no other ideas; hence we've resolved 369. Now the 3021. I establish that the number 21 is 7 by 3 and excuse it as an excess error of some triangular stationed vector in this lost part of +3 to get at 10 by 51 less 41.

Reducing the 3021 to 3020 to get a better value of divisible use, the construction of our model being entirely lost to strength of its mass and volume at length of a meaning that is not two-dimensional as we have stated it, I withdraw the number 1 increment to signify dimension as we have elsewhere above; and also withdraw results of any .5 values to signify binary domains in results. And the only good divisibles in the litany under 10, are at 1, 2, 4 and 5 values, resulting in clean and unencumbered values, which we could believe as perfect and magical. But this progression is valuable in itself. It requires 4 clean values of use under the perfect value of 10 and so it could signify the usages of the sides each of 9 and revolving around the object of 9 progressing to 10 by the dimensional usage of binary .5 to 1 in our scale, with 1 being the fixed point up to the value of 9; and also omitting the other usages of 3, its double 6, also 7, 8, and 9. The universe is thereby clean to contemplate that the 1, 2, 4, 5 progression solution results in the values of 3020, 1510, 755, 604. Which all together equate a value of quantity that is 5889, but of its magic has an interesting story. Its front and end values in digits differ by 4 which are the 4 square basis lines. And the interiors are two 8's which account to 16 and may be cubed to make the dimensional quadrant of the square basis. So possibly there is some logic and code alike, making magic in these sequences. And if there is not, I account it finished in story by me.)

The Magic Tables, Seals & Characters of the Planets, their Intelligences & Spirits

The Table of the Sun in his Compass

6	32	3	34	35	1
7	11	27	28	8	30
19	14	16	15	23	24
18	20	22	21	17	13
25	29	10	9	26	12
36	5	33	[illegible]	2	31

The same in Hebrew

ו	לב	ג	לד	לה	א
ז	יא	כז	כח	ח	ל
יט	יד	יו	יה	כג	כד
יח	כ	כב	כא	יז	יג
כה	כט	י	ט	כו	יב
לו	ה	לג	ד	ב	לא

The Character of the Seal of the Sun

His Intelligence

His Spirit

The Table of Venus in her Compass

22	47	16	41	10	35	4
5	23	48	17	42	11	29
30	6	24	49	18	36	12
13	31	7	25	43	19	37
38	14	32	1	26	44	20
21	39	8	33	2	27	45
46	15	40	9	34	3	28

In Hebrew

כב	מז	יו	מא	י	לה	ד
ה	כג	מח	יז	מב	יא	כט
ל	ו	כד	מט	יח	לו	יב
יג	לא	ז	כה	מג	יט	לז
לח	יד	לב	א	כו	מד	כ
כא	לט	ח	לג	ב	כז	מה
מו	טו	מ	ט	לד	ג	כח

The Seal of Venus

Her Intelligence

Her Spirit

Her Intelligences

F. Barrett Del. Pub. by Lackington & Allen Griffith Sculp.

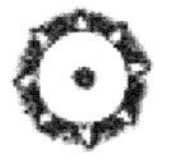

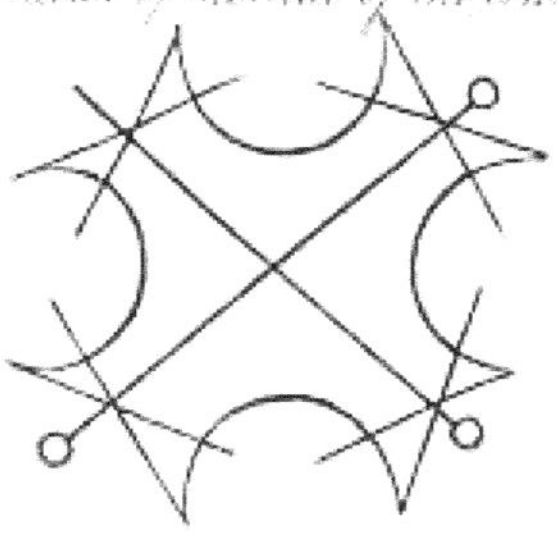

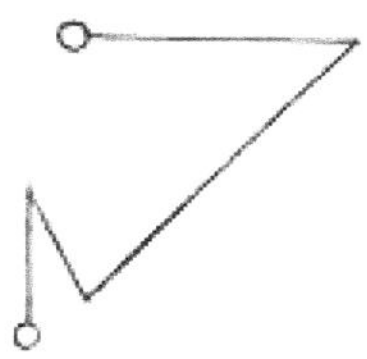

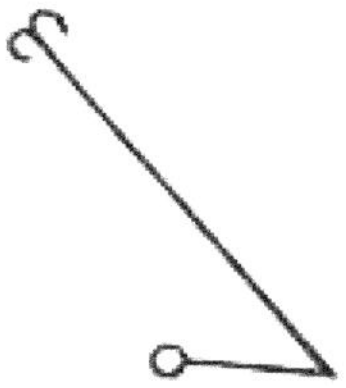

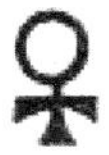

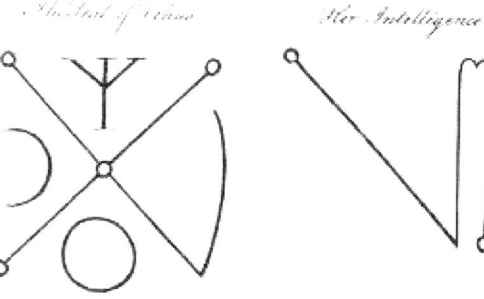

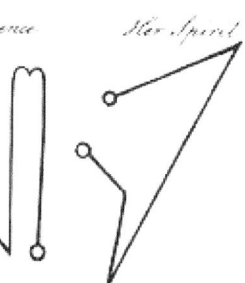

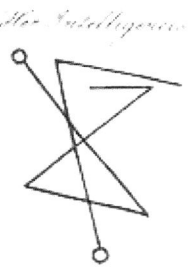

The Magick Tables, Seal & Characters of the Planet, their Intelligence & Spirit.

The Table of Mercury in his Compass.   The same in Hebrew.

8	58	59	5	4	62	63	1
49	15	14	52	53	11	10	56
41	23	22	44	45	19	18	48
32	34	35	29	28	38	39	25
40	26	27	37	36	30	31	33
17	47	46	20	21	43	42	24
9	55	54	12	13	51	50	16
64	2	3	61	60	6	7	57

ח	נח	סג	א	ד	נט	סב	ח
מט	יה	יד	נב	נג	יא	י	נו
מא	כג	כב	מד	מה	יט	יח	מח
לב	לד	לה	כט	כח	לח	לט	כה
מ	כו	כז	לז	לו	ל	לא	לג
יז	מז	מו	כ	כא	מג	מב	כד
ט	נה	נד	יב	יג	נא	נ	טז
סד	ב	ג	סא	ס	ו	ז	נז

☿

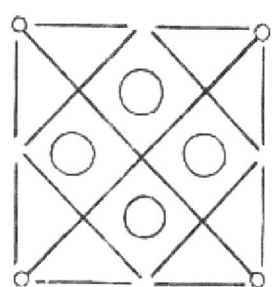

The Seal & Character of Mercury

The Character of the Intelligence of Mercury

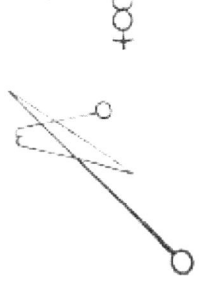

The Character of the Spirit of Mercury

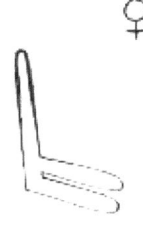

The Magic Tables, Seals & Characters of the Planets, their Intelligences & Spirits

The Table of the Moon in her compass   Table of the ☽ in Hebrew Notes

(magic square of the Moon — numerical)   *(magic square of the Moon — Hebrew letters)*

Seal of the Moon   Character of the Spirit of ☽

of the Spirit of Spirits   of the Intelligence of the Intelligence
of the Moon   of the Moon

Names answering to the Numbers of Saturn. ♄

Numbers.	Divine Names.	Divine Names in Hebrew.
3	Ab	אב
9	Hod	הד
15	Jah	יה
15	Hod	חיד
45	Jehovah extended	יזרהאואהא
45	Agiel, the Intelligence of Saturn	אגיאל
45	Zazel, the Spirit of Saturn	זאזל

Names answering to the Numbers of Jupiter. ♃

4	Aba	אבא
16		הוה
16		אהי
34	El Ab	אלאב
136	Johphiel, the Intelligence of Jupiter	יהפיאל
136	Hismæl, the Spirit of Jupiter	הסמאל

Names answering to the Numbers of Mars. ♂

5	He, the letter of the holy name	ה
25		יהי
65	Adonai	p. 147 אדני
325	Graphiel, the Intelligence of Mars	גראפיאל
325	Barzabel, the Spirit of Mars	ברצאבאל

Names answering to the Numbers of the Sun. ☉

6	Vau, the letter of the holy name	ו
6	He extended, the letter of the holy name	הא
36	Eloh	אלה
111	Nachiel, the Intelligence of the Sun	נכיאל
666	Sorath, the Spirit of the Sun	סורה

Names answering to the Numbers of Venus. ♀

7	Aha	אהא
49	Hagiel, the Intelligence of Venus	הגיאל
175	Kedemel, the Spirit of Venus	קדמאל
1225	Bne Seraphim, the Intelligence of Venus	בני שרפים

Names answering to the Numbers of Mercury. ☿

8	Asboga, eight extended	אזבגה
64	Din	דין
64	Doni	דני
260	Tiriel, the Intelligence of Mercury	טיריאל
2080	Taphthartharath, the Spirit of Mercury	תפתרתרת

Names answering to the Numbers of the Moon. ☽

9	Hod	הד
81	Elim	אלים
369	Hasmodai, the Spirit of the Moon	השמודאי
3321	Schedbarschemoth Schartathan, the Spirit of the Spirits of the Moon	שדבשהמעהשרתתי
3321	Malcha betharsisim hed beruah schehalim, the Intelligence of the Intelligences of the Moon	קלכאבתדשימערברוחשהקים

Chapter 29
Of the Observation of the Celestials
Necessary in Every MAGICAL WORK

Every Natural Virtue works things far more wonderful when it is not only compounded of a Natural Proportion, but also is informed by a choice observation of the Celestials Opportune to this (viz. when the Celestial Power is most strong to that effect which we desire and also helped by many Celestials), by subjecting Inferiors to the Celestials, as proper females, to be made fruitful by their males.

Also, in every work, there are to be observed the situation, motion and aspect of the Stars and Planets, in Signs and Degrees and how all these stand in reference to the Length and Latitude of the Climate. For by this are varied the Qualities of the Angles, which the Rays of the Celestial Bodies upon the Figure of the Thing, describe, according to which Celestial Virtues are infused.

So when you are working anything which belongs to any Planet, you must place it in its Dignities Fortunate and Powerful and Ruling in the Day Hour and in the Figure of the Heavens. Neither must you expect the Signification of the Work to be Powerful, but you must observe the Moon opportunely directed to this. For you shall do nothing without the Assistance of the Moon. And if you have more patterns of your work, observe them all, being most powerful and looking upon one another with a friendly aspect. And if you cannot have such aspects, it will be convenient at least that you take them angular. But you shall take the Moon either when she looks upon both, or is joined to one and looks upon the other, or when she passes from the conjunction or aspect of one, to the conjunction or aspect of the other. For that, I conceive, I must in no wise be omitted.

Also, you shall in every work observe Mercury, for he is a Messenger between the Higher Gods and the Infernal Gods. When he goes to the Good, he increases their Goodness; when he goes to the Bad, he has influence on their Wickedness. We call it Unfortunate Sign or Planet, when it is, by the Aspect of Saturn or Mars, especially, opposite or quadrant, for these are the aspects of Enmity.

But a conjunction, a trine and a sextile aspect are of Friendship. Between these there is a greater conjunction. But yet if you do already behold it through a trine and the Planet be received, it is accounted as already conjoined. Now all Planets are afraid of the Conjunction of the Sun, rejoicing in the trine and sextile aspect thereof.

Chapter 30
When the PLANETS are of Most Powerful Influence

Now we shall have the Planets powerful when they are ruling in a House or in Exaltation or Triplicity or Term or Face, without combustion of what is direct in the Figure of the Heavens, viz. when they are in angles, especially of the Rising or Tenth, or in Houses presently succeeding, or in their Delights.

But we must take heed that they are not in the bounds or under the Dominion of Saturn or Mars, lest they be in Dark Degrees, in Pits or Vacuities. You shall observe that the Angles of the Ascendant and Tenth and Seventh be Fortunate; as also the Lord of the Ascendant and Place of the sun and Moon; and Place of the Past of Fortune and the Lord thereof, the Lord of the Foregoing Conjunction and Prevention.

But that they of the Malignant Planet fall Unfortunate; unless happily they be significations of your work or can be of any advantage to you or in your revolution or birth they had the predominance, for then they are not at all to be depressed.

Now we shall have the Moon powerful if she be in her House or Exaltation or Triplicity or Face or in Degree Convenient for the desired work; and if it has a Mansion of these Twenty-Eight, suitable to itself and the work, let her not in the way be burnt up, nor slow in course; let her not be in the Eclipse or burnt by the Sun; let her not descend in the Southern Latitude, when the she goes out of the burning. Neither, let her be opposite to the Sun, nor deprived of Light. Let her not be hindered by Mars or Saturn.

Chapter 31
Observations on the FIXED STARS and their Names and Natures

There is the like consideration to be had in all things concerning the Fixed Stars. Know this, that all the Fixed Stars are of the Signification and Nature of the Seven Planets; but some are of the Nature of One Planet and some of Two Planets. Hence, as often as any planet is joined with any of the Fixed Stars of its own Nature, the Signification of that Star is made more powerful and the Nature of the Planet augmented.

But if it be a Star of Two Natures, the Nature of that which shall be the stronger with it, shall overcome in Signification; as for example, if it be of the Nature of Mars and Venus, if Mars shall be the stronger two with it, the Nature of Mars shall overcome; but if Venus be the stronger two with it, the Nature of Venus shall overcome.

Now the Natures of Fixed Stars are discovered by their Colors, as they agree with Certain Planets and are ascribed to them. Now the Colors of the Planets are these ~ of Saturn, blue and leaden and shining with blue lead; of Jupiter, citrine, near to a paleness and clear with this pale citrine; of Mars, red and fiery; of the Sun, yellow and when it rises, red, but afterwards, glittering; of Venus, white and shining, but white in the morning and reddish in the evening; of Mercury, glittering; of the Moon, fair.

Know, also, that of the Fixed Stars, by how much the greater and brighter and apparent they are, so much the greater and stronger is the Signification. Such are those Stars called by the Astrologers of the First and Second Magnitude.

I will tell you some of these which are more potent to this Faculty, viz.

~ the Navel of Andromeda, in the Twenty-Second Degrees of Aries of the Nature of Venus and Mercury; some call it jovial and saturnine;

~ the Head of Algol, in the Eighteenth Degree of Taurus, of the Nature of Saturn and Jupiter;

~ the Pleiades are also in the Twenty-Second Degree, a Lunary Star by Nature and Complexion Martial;

~ also Aldeboram, in the Third Degree of Gemini, is of the Nature of Mars and Complexion of Venus;

~ but Hermes places this in the Twenty-Fifth Degree of Aries;

~ the Goat Star, in the Thirteenth Degree of Gemini, is of the Nature of Jupiter and Saturn;

~ the Great Dog Star is the Seventeenth Degree of Cancer and Venereal;

~ the Little Dog Star is in the Seventeenth Degree of Cancer and Venereal and is of the Nature of Mercury and Complexion of Mars;

~ the King Star, which is called the Heart of the Lion, is the Twenty-First Degree of Leo and the Nature of Jupiter and Mars;

~ the Tail of the Great Bear is in the Nineteenth Degree of Virgo and is Venereal and Lunary;

~ the Star which is called the Right Wing of the Crow, is in the Seventh Degree of Libra;

~ and in the Thirteenth Degree of Libra, is the Left Wing of the Crow and it is both the Nature of Saturn and Mars;

~ the Star called Spica, is in the Sixteenth Degree of Libra and is Venereal and Mercurial;

~ in the Seventeenth Degree of Libra is Alcameth of the Nature of Mars and Jupiter;

~ but of Alcameth, when the Sun's aspect is full towards it, it remains the Nature of Jupiter, but when the Sun's aspect is contrary it, it remains the Nature of Mars;

~ Elepheia is in the Fourth Degree of Scorpio and is of the Nature of Venus and Mars;

~ the Heart of the Scorpion is in the Third Degree of Sagittarius and is of the Nature of Mars and Jupiter;

~ the Falling Vulture is in the Seventh Degree of Capricorn and it is Temperate, Mercurial and Venereal;

~ the Tail of Capricorn is in the Sixteenth Degree of Aquarius and is of the Nature of Saturn and Mercury;

~ the Star called the Shoulder of the Horse, which is in the Third Degree of Pisces, is of the Nature of Jupiter and Mars;

    And it shall be a General Rule for you to expect the proper gifts of the Stars while they Rule; to be prevented of them; they being Unfortunate, as is above shown.
    For Celestial Bodies, inasmuch as they are affected Fortunately or Unfortunately, so much do they affect us, our works and those things which we use, fortunately or unhappily.
    And although many effects proceed from the Fixed Stars, yet they are attributed to the Planets as because being more near to us and more distinct and known, so because they execute whatever the Superior Stars communicate to them.

Chapter 32
Of the SUN and MOON
And their MAGICAL Considerations

The Sun and Moon have obtained the Administration of Ruling the Heavens and all Bodies under the Heavens.

The Sun is the Lord of All Elementary Virtues and the Moon, by Virtue of the Sun, is Mistress of Generation, increase or decrease.

Albumasar says, that by the Sun and Moon, Life is infused into all things, which Orpheus calls the enlivening Eyes of Heaven.

The Sun gives light to all things of itself and gives it plentifully, not only to all things in Heaven and Air, but Earth and Deep.

Whatever good we have, Jamblicus says, we have it from the Sun alone, or from it through other things.

Heraclitus calls the Sun, the Fountain of Celestial Light and many of the Platonists placed the Soul of the World chiefly in the Sun, as that which, filling the whole Globe of the Sun, does send forth its rays on all sides, as it were a Spirit through all things, distributing life, sense and motion to the Universe.

Hence the Ancient Naturalists called the Sun, the Very Heart of Heaven and the Chaldeans put it as the Middle of the Planets.

The Egyptians also placed it in the Middle of the World, viz. between the Two Fives of the World; that is, above the Sun, they place Five Planets and under the Sun, they place the Moon and Four Elements. For it is, amongst the other Stars, the Image and Statue of the Great Prince of Both Worlds, viz. Terrestrial and Celestial.

The Sun is the True Light and the most exact Image of God Himself, whose essence resembles the Father; light, the Son; heat, the Holy Ghost. So that the Platonists have nothing to hold forth the Divine Essence more manifestly by than this.

The Sun disposes even the very Spirit and Mind of Man, which Homer says and is approved by Aristotle, that there are in the mind, such like motions as the Sun, the Prince and Moderator of the Planets, brings to us every day.

But the Moon, being the nearest to the Earth, is the Receptacle, or Receptor, of all the Heavenly Influences, by the swiftness of her course and is joined to the Sun and the other Planets and Stars, every Month. And receiving the beams and influences of all other Planets and Stars, as a conception, the Moon brings all these lights and influences forth to be received into the Inferior World, as being next to itself – for all the Stars have influence on it, its place being that of the last Receiver. By which place, it may afterwards communicate the Influence of all the Superiors to these Inferiors and pours forth upon the Earth. And it more manifestly disposes the Influences and Lights to these Inferiors than the others may in their greater disposition to (astral or celestial or ethereal) things.

Therefore, her motion is to be observed before the others, as the parent of all conceptions, which it diversely issues forth in these Inferiors according to the diverse complexity, motion, situation and different aspects to the Planets and other Stars.

And though it receives Powers from all the Stars, yet especially from the Sun, as often as it is in conjunction with the same, it is replenished with vivifying Virtue; and, according to the aspect thereof, it borrows its complexion (composition).

From it, the Heavenly Bodies begin that Series of Things, which Plato calls, the Gold Chain, by which everything and every cause, being linked one to another, does depend on the Superior, even until it may be brought to the Supreme Cause of all, from which all things depend.

Hence it is, that, without the Moon intermediating, we cannot at any time attract the Power of the Superiors. Therefore, to obtain the Virtue of any Star, take the Stone and Herb of that Planet, when the Moon Fortunately comes under, or has a Good Aspect on, that Star.

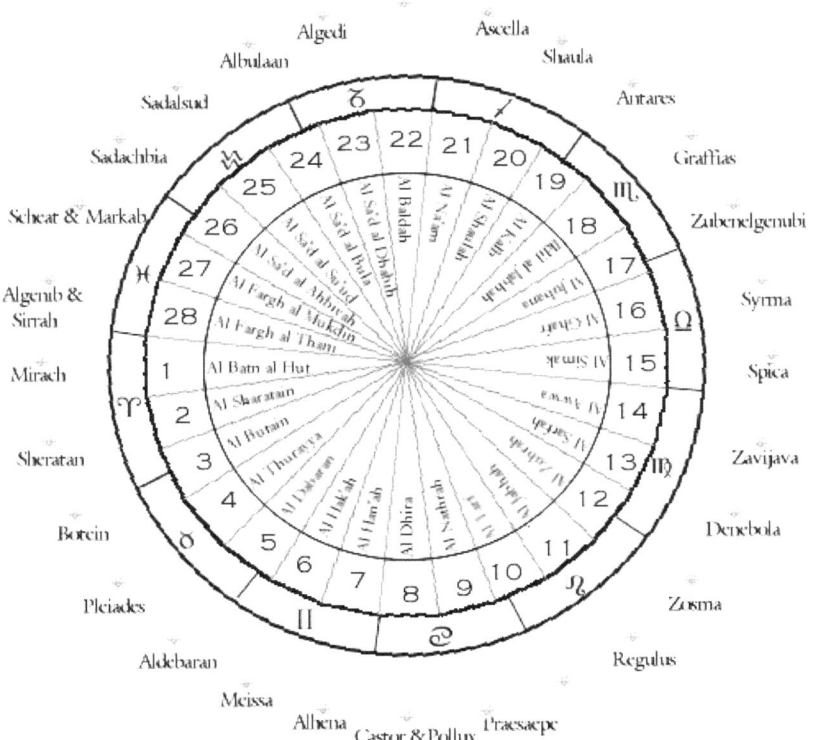

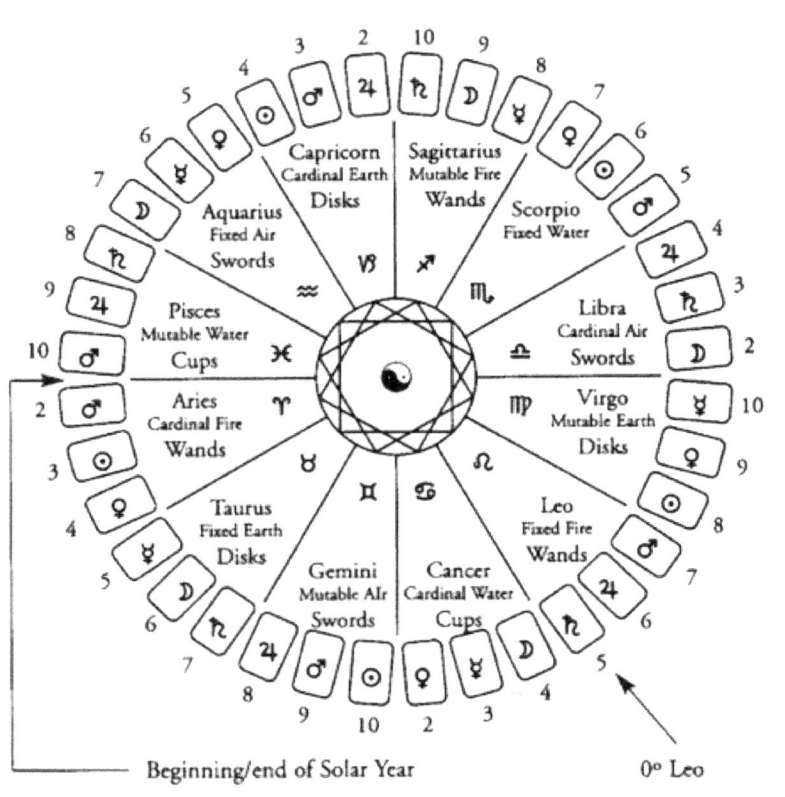

Chapter 33
Of the 28 MANSIONS OF THE MOON and their VIRTUES

And seeing the Moon measures the whole space of the Zodiac in the time of Twenty-Eight Days, hence it is that the Wisemen of the Indians and most of the Ancient Astrologers have granted Twenty-Eight Mansions to the Moon, which, being Fixed in the Eighth Sphere, do enjoy (as Alpharus says) diverse names and properties, from the Various Signs and Stars which are contained in them; through which, while the Moon wanders, it obtains many other Powers and Virtues. But every one of these Mansions, according to the opinion of Abraham, contained Twelve Degrees and Fifty-One Minutes and almost Twenty-Six Seconds, whose Names, and also their Beginnings in the Zodiac, of the Eighth Sphere, are these.

The FIRST MANSION is called ALNATH.

That is; the Horns of Aries.
His Beginning is from the Head of Aries, of the Eighth Sphere. It causes discords and journeys.

The SECOND MANSION is called ALLOTHAIM, or ALBOCHAN.

That is; the Belly of Aries.
And his Beginning is from the Twelfth Degree of the same Sign, Fifty-One Minutes, Twenty-Two Seconds complete. It conduces to the finding of Treasures and to the retaining captives.

The THIRD MANSION is called ACHAOMAZON, or ATHORAY.

That is, Showering, or Pleiades.
His Beginning is from the Twenty-Fifth Degree of Aries complete, Forty-Two Minutes and Fifty-One Seconds. It is profitable to sailors, huntsmen and alchemists.

The FOURTH MANSION is called ALDEBARAM, or ALDELAMEN.

That is, the Eye or Head of Taurus.
His Beginning is from the Eighth Degree of Taurus, Thirty-Four Minutes and Seventeen Seconds of the same, Taurus being excluded. It causes the destruction and hindrances of buildings, fountains, wells, gold mines, the flight of creeping things and begets discord.

The FIFTH MANSION is called ALCHATAY, or ALBACHAY.

The Beginning of it is after the Twenty-First Degree of Taurus, Twenty-Five Minutes, Forty Seconds. It helps to the return from a journey, to the instruction of scholars. It confirms edifices and gives health and goodwill.

The SIXTH MANSION is called ATHANNA, or ALCHAYA.

That is, the Little Star of Great Light.
His Beginning is after the Fourth Degree of Gemini, Seventeen Minutes and Nine Seconds. It conduces to hunting and besieging towns and to the revenge of Princes. It destroys harvests and fruits and hinders the operation of the Physician.

The SEVENTH MANSION is called ALDIMICAH, or ALARZACH.

That is, the Arm of Gemini.
It begins from the Seventeenth Degree of Gemini, Eight Minutes and Thirty-Four Seconds and lasts even to the end of the Sign. It confirms gain and friendship. It is profitable to lovers and destroys magistracies. And so is one quarter of the Heaven completed in these Seven Mansions and in the like Order and Number of Degrees, Minutes and Seconds the remaining Mansion, in every quarter, have their several beginnings namely, so that in the First Sign of this Quarter Three Mansions take their Beginnings.

In the other Two Signs, Two Mansions (take their Beginnings) in each. Therefore, the Seven Following Mansions begin with Cancer.

The EIGHTH MANSION is called ALMAZA ANATRACHYA.

That is, Misty or Cloudy.
It causes love, friendship and society of fellow travelers. It drives away mice and afflicts captives, confirming their imprisonment.

The NINTH MANSION is called ARCHAAM, or ARCAPH.

That is, the Eye of the Lion.
It hinders harvest and travelers and puts discord between men.

The TENTH MANSION is called ALGELIOCHE, or ALBGEBH.

That is, the Neck or Forehead of Leo.
It strengthens buildings, promotes love, benevolence and help against enemies.

The ELEVENTH MANSION is called AZOBRA, or ARDAF.

That is, the Hair of the Lion's Head.
It is good for voyages and gain by merchandise and for redemption of captives.

The TWELFTH MANSION is called ALZARPHA, or AZARPHA.

That is, the Tail of Leo.
It gives prosperity to harvest and plantations, but hinders seamen and is good for the bettering of servants, captives and companions.

The THIRTEENTH MANSION is called ALHAIRE.

    That is, Dog Stars, or the Wings of Virgo.
    It is prevalent for benevolence, gain, voyages, harvests and freedom of captives.

The FOURTEENTH MANSION is called ACHURETH, or ARIMET; by others, AZIMETH, or ATHUMECH, or ALCHEYMECH.

    That is, the Spike of Virgo, or Flying Spike.
    It causes the love of married folks; it cures the sick, is profitable to sailors, but hinders journeys by land; and in these, the Second Quarter of the Heaven is completed.

THE OTHER SEVEN FOLLOW.
The First of which Begins in the Head of Libra.

The FIFTEENTH MANSION is called AGRAPHA, or ALGRAPHA.

    That is, Covered, or Covered Flying.
    It is profitable for extracting treasures, for digging of pits; it assists divorce, discord and destruction of houses and enemies and hinders travelers.

The SIXTEENTH MANSION is called AZUBENE, or AHUBENE.

    That is, the Horns of Scorpio.
    It hinders journeys and wedlock, harvest and merchandise; it prevails for redemption of captives.

The SEVENTEENTH MANSION is called ALCHIL.

    That is, the Crown of Scorpio.
    It betters a bad fortune, makes love durable, strengthens buildings and helps seamen.

The EIGHTEENTH MANSION is called ALCHAS, or ALTOB.

    That is, the Heart of Scorpio.
    It causes, discord, sedition, conspiracy against Princes and Mighty Ones and revenge from enemies. But it frees captives and helps edifices.

The NINETEENTH MANSION is called ALLATHA, or ACHALA; by others, HYCULA, or AXALA.

    That is, the Tail of Scorpio.
    It helps in besieging of cities and taking of towns and in the driving of men from their places and for the destruction of seamen and perdition of captives.

The TWENTIETH MANSION is called ABNAHAYA.

    That is, a Beam.
    It helps for the taming of wild beasts, for strengthening of prisons; it destroys the wealth of societies. It compels a man to come to a certain place.

The TWENTY-FIRST is called ABEDA, or ALBELDACH.

    That is, a Desert.
    It is good for harvest, gain, buildings and travelers and causes divorce; and in this, is the third quarter of Heaven completed.

There remains the SEVEN LAST MANSIONS,
completing the Last Quarter of Heaven.
The First of which, Begins from the Head of Capricorn.

The TWENTY-SECOND MANSION is called SADAHACHA,
or ZODEBOLUCH, or ZANDELDENA.

> That is, a Pastor.
> It promotes the Flight of Servants and Captives, that they may escape and helps the curing of diseases.

The TWENTY-THIRD is called ZABADOLA, or ZOBRACH.

> That is, Swallowing.
> It is for divorce, liberty of captives and health to the sick.

The TWENTY-FOURTH is called SADABATH, or CHADEZOAD.

> That is, the Star of Fortune.
> It is prevalent for the benevolence of married people, for the victory of soldiers. It hurts the execution of government and prevents its being exercised.

The TWENTY-FIFTH is called SADALABRA, or SADALACHIA.

> That is, a Butterfly, or a Spreading Forth.
> It favors besieging and revenge. It destroys enemies and causes divorce; confirms prisons and buildings, hastens messengers. It conduces to spells against copulation (or intercourse), and so binds every member of man that is cannot perform its duty.

The TWENTY-SIXTH is called ALPHARG,
or PHRAGAL MOCADEN.

    That is, the First Drawing.
    It causes union, health of captives, destroys building and prisons.

The TWENTY-SEVENTH is called ALCHARA ALYHALGALMOAD.

    Or, the Second Drawing.
    It increases harvests, revenues, gain and heals infirmities; but hinders buildings, prolongs prisons, causes danger to seamen and helps to infer mischiefs on whom you shall please.

The TWENTY-EIGHTH is called ALBOTHAM, or ATCHALCY.

    That is, Pisces.
    It increases harvest and merchandise. It secures travelers through dangerous places. It makes for the joy of married people, but it strengthens prisons and causes loss of treasures.

    AND in these TWENTY-EIGHT MANSIONS, lie hid many SECRETS of WISDOM of the ANCIENTS, by which they wrought WONDERS on ALL THINGS which are UNDER the CIRCLE of the MOON. And they attributed to every MANSION, His Resemblances, Images, Seals and His President Intelligences and worked by the Virtue of them after different manners.

The Twenty-Eight Mansions of the Moon (Arabic System)

	Lunar Mansion Name	Arabic Name	Meaning	Position (to the nearest minute)	Constellation	Planetary Ruler
1	Alnath	Al Sharatain	The Two Signs	0	Aries	Sun
2	Albotain	Al Butain	The Belly	12.51	Aries	Moon
3	Azoraya	Al Thurayya	The Many Little Ones	25.43	Aries	Mars
4	Aldebaran	Al Dabaran	The Follower	8.34	Taurus	Mercury
5	Almices	Al HaKah	A White Spot	21.26	Taurus	Jupiter
6	Athaya	Al Hanah	A Brand or Mark	4.17	Gemini	Venus
7	Aldirah	Al Dhira	The Forearm	17.9	Gemini	Saturn
8	Annathra	Al Nathrah	The Gap or Crib	0	Cancer	Sun
9	Atarf	Al Tarf	The Glance of the Lion's Eye	12.51	Cancer	Moon
10	Algebha	Al Jabhah	The Forehead	25.43	Cancer	Mars
11	Azobra	Al Zubrah	The Mane of the Lion	8.34	Leo	Mercury
12	Acarfa	Al Sarfah	The Changer of the Weather	21.26	Leo	Jupiter
13	Alahue	Al Awwa	The Barker	4.17	Virgo	Venus
14	Azimech	Al Simak	The Unarmed	17.9	Virgo	Saturn
15	Argafra	Al Ghafr	The Covering	0	Libra	Sun
16	Azubene	Al Jubana	The Claws	12.51	Libra	Moon
17	Alichil	Iklil Al Jabhah	The Crown of the Forehead	25.43	Libra	Mars
18	Alcalb	Al Kalb	The Heart	8.34	Scorpio	Mercury
19	Exaula	Al Shaula	The Sting	21.26	Scorpio	Jupiter
20	Nahaym	Al Na'am	The Ostriches	4.17	Sagittarius	Venus
21	Elbelda	Al Baldah	The City or District	17.9	Sagittarius	Saturn
22	Caadaldeba	Al Sad Al Dhabih	The Lucky One of The Slaughterers	0	Capricorn	Sun
23	Caadebolach	Al Sad Al Bulah	The Good Fortune of The Swallower	12.51	Capricorn	Moon
24	Caadacohot	Al Sad Al	The Luckiest of the Su'udLucky	25.43	Capricorn	Mars
25	Caadalhacbia	Al Sad Al Ahbiya	The Lucky Star of Hidden Things	8.34	Aquarius	Mercury
26	Almiquedam	Al Farch Al Mukdm	The Forespout of the Water bucket	21.26	Aquarius	Jupiter
27	Algarfalmuehra	Al Fargh Al Thani	The Lower Spout of the Water bucket	4.17	Pisces	Venus
28	Arrexhe	Al Batn al Hut	The Belly of the Fish	17.9	Pisces	Saturn

The Twenty-Eight Mansions of the Moon (Oriental System)

Four Symbols (四象)	Mansion (宿)			
	Number	Name (pinyin)	Translation	Determinative star
Azure Dragon of the East (東方青龍) Spring	1	角 (Jiǎo)	Horn	α Vir
	2	亢 (Kàng)	Neck	κ Vir
	3	氐 (Dǐ)	Root	α Lib
	4	房 (Fáng)	Room	π Sco
	5	心 (Xīn)	Heart	σ Sco
	6	尾 (Wěi)	Tail	μ Sco
	7	箕 (Jī)	Winnowing Basket	γ Sgr
Black Tortoise of the North (北方玄武) Winter	8	斗 (Dǒu)	(Southern) Dipper	φ Sgr
	9	牛 (Niú)	Ox	β Cap
	10	女 (Nǚ)	Girl	ε Aqr
	11	虛 (Xū)	Emptiness	β Aqr
	12	危 (Wēi)	Rooftop	α Aqr
	13	室 (Shì)	Encampment	α Peg
	14	壁 (Bì)	Wall	γ Peg
White Tiger of the West (西方白虎) Fall	15	奎 (Kuí)	Legs	η And
	16	婁 (Lóu)	Bond	β Ari
	17	胃 (Wèi)	Stomach	35 Ari
	18	昴 (Mǎo)	Hairy Head	17 Tau
	19	畢 (Bì)	Net	ε Tau
	20	觜 (Zī)	Turtle Beak	λ Ori
	21	參 (Shēn)	Three Stars	ζ Ori
Vermilion Bird of the South (南方朱雀) Summer	22	井 (Jǐng)	Well	μ Gem
	23	鬼 (Guǐ)	Ghost	θ Cnc
	24	柳 (Liǔ)	Willow	δ Hya
	25	星 (Xīng)	Star	α Hya
	26	張 (Zhāng)	Extended Net	υ¹ Hya
	27	翼 (Yì)	Wings	α Crt
	28	軫 (Zhěn)	Chariot	γ Crv

Chapter 34
How some ARTIFICIAL THINGS
(as Images, Seals and such like)
may obtain some VIRTUES from the CELESTIAL BODIES

So great is the extent, power and efficacy of the Celestial Bodies, that not only Natural Things, but also Artificial, when they are rightly exposed to those above, do presently suffer by that most potent agent and obtain a wonderful Life.

The Magicians affirm, that not only by the mixture and application of Natural Things, but also in Images, Seals, Rings, Glasses and some other Instruments, being opportunely framed under a certain Constellation, some Celestial Illustrations may be taken and some wonderful thing may be received.

For the Beams of the Celestial Bodies, being animated, living, sensual and bringing along with them admirable gifts and a most violent power, do, even in a moment and at the first touch, imprint wonderful powers in the Images, though their Matter be less capable.

Yet they bestow more powerful Virtues on the Images if they be framed not of any, but of a certain Matter, namely, whose Natural, but also Specific Virtue, is agreeable with the work; and the Figure of the Image is like to the Celestial.

For such an Image, both in regard to the Matter naturally congruous to the operation and Celestial Influence and also for its Figure being like to the Heavenly One, is best prepared to receive the operations and powers of the Celestial Bodies and Figures and instantly receives the Heavenly Gift into itself, though it constantly works on another thing and other things yield obedience to it.

Geomantic Characters

Chapter 35
Of the Images of the Zodiac
What Virtues They Receiver from the Stars, Being Engraved

But the Celestial Images, according to whose likeness Images of this kind are framed, are many in the Heavens, some visible, arid, conspicuous and others only imaginary, conceived and set down by the Egyptians, Indians and Chaldeans. And their parts are so ordered, that even the figures of some of them are distinguished from others. For this reason, they place in the Circle of the Zodiac, Twelve General Images, according to the Number of the Signs.

Of these, they constituting Aries, Leo and Sagittarius (for the Fiery and Oriental Triplicity), report that it is profitable against fevers, Palsy, dropsy, gout and all cold and phlegmatic infirmities. And that it makes him who carries it to be acceptable, eloquent, ingenious and honorable; because they are the Houses of Mars, Sol and Jupiter.

They made, also, the Image of a Lion against melancholy fantasies, dropsy, plague and fevers and to expel diseases. At the Hour of the Sun, the First Degree of the Sign Leo ascending, which is the Face and decant of Jupiter. But against the "stone" and diseases of the reins (kidneys) and against hurts of the Beasts – they made the Image, when Sol, in the Heart of the Lion, obtained the Midst of Heaven.

And again, because Gemini, Libra and Aquarius, do constitute the aerial and Occidental Triplicity and are the Houses of Mercury, Venus and Saturn, they are said to put to flight diseases, to conduce to friendship and concord, to prevail against melancholy and to cause health.

And they report that Aquarius especially frees from the Quartan (Fever). Also, they report that Cancer, Scorpio and Pisces, because they constitute the watery and Northern Triplicity, do prevail against hot and dry fevers and also against the hectic and all choleric Passions. But as for Scorpio, because among the "members" it respects the privy parts, it does provoke to Lust.

But these did frame it for this purpose, his Third Face ascending, which belongs to Venus and they made the same, against Serpents and Scorpions, Poisons and Evil Spirits, his Second Face ascending, which is the Face of the Sun and decant of Jupiter.

And they report that it makes him who carries it, Wise, of a Good Color. And they say that the Image of Cancer is most efficacious against Serpents and Poison, when Sol and Luna are in conjunction in it and ascend in the First and Third Face. For this is the Face of Venus and the decant of Luna. But the Second Face of Luna, is the decant of Jupiter. They report, also, that Serpents are tormented when the Sun is in Cancer. Also, that Taurus, Virgo and Capricorn, because they constitute the Earthly and southern Triplicity, do cure hot infirmities and prevail against the Synocal (?) Fever.

It makes those who carry it, grateful, acceptable, eloquent, devout and religious, because they are the Houses of Venus, Mars and Saturn, Capricorn, also is reported to keep men in safety and also places in security, because it is the Exaltation of Mars.

Chapter 36
Of the Images of Saturn

But now let us examine what Images they did attribute to the Planets. Although of these things, very large volumes have been written by the Ancient Wisemen, so that there is no need to declare them here.

Notwithstanding, I will recite a few of them, for they made, from the operations of Saturn, Saturn ascending in a Stone, which is called the Loadstone, the Image of a Man, having the countenance of a Hart and Camel's Feet; and sitting upon a Chair; or else a Dragon, holding in his right Hand a Scythe, in his left Hand, a Dart, which Image they hoped would be profitable for prolongation of life.

For Albumasar, in his book, Sadar, proves that Saturn conduces to the Prolongation of Life. Where also, he says that of a certain region of India, being subject to Saturn there, men are of a very long life and die not unless by extreme old age.

They made also, an Image of Saturn, for length of days, in a Sapphire, at the hour of Saturn, Saturn ascending or fortunately constituted; whose figure was an old Man sitting upon a high Chair, having his Hands lifted up above his Head and in them, holding a Fish or Sickle and under his Feet, a Bunch of Grapes, his Head covered with a black of dusky colored Cloth and all his Garments black or dark.

They also make this same Image against the Stone and Disease of the Kidneys, viz. in the Hour of Saturn, Saturn ascending with the Third Face of Aquarius.

They made also, from the operations of Saturn, an Image for the increasing of Power, Saturn ascending in Capricorn; the form of which was an old Man leaning on a Staff, having in his Hand, a Crooked Sickle and clothed in black.

They also made, in an Image of Melted Copper, Saturn ascending in his rising, viz. in the First Degree of Aries, or the First Degree of Capricorn, which Image they affirm to speak with a Man's voice.

They made also, from the operations of Saturn and also Mercury, an Image of Cast Metal, like a beautiful Man, which, they said, would foretell things to come and made it on the Day of Mercury, on the Third Hour of Saturn, the Sign of Gemini ascending, being the House of Mercury, signifying Prophets; Saturn and Mercury being in conjunction in Aquarius, in the Ninth House of Heaven, which is also called (the very Throne of) God.

Moreover, let Saturn have a trine aspect on the Ascendant and the Moon in like manner and the Sun, will have an aspect on the place of conjunction. Venus, obtaining some angle, may be powerful and occidental. Let Mars be combust by the Sun, but let it not have an aspect on Saturn and Mercury, for they said that the Splendor of the Powers of these Stars was diffused upon this Image and it did speak with men and declare those things which are profitable for them.

Chapter 37
Of the Images of Jupiter

From the Operations of Jupiter they made, for Prolongation of Life, an Image in the Hour of Jupiter, Jupiter being in his Exaltation Fortunately ascending, in a clear and White Stone, whose figure was a Man crowned and clothed with Garments of a Saffron color, riding upon an Eagle or Dragon, having in his right Hand a Dart, about, as it were, to strike it into the Head of the same Eagle or Dragon.

They made, also, another Image of Jupiter, at the same convenient season, in a white and clear Stone, especially in Crystal. And it was a naked Man crowned, having both his Hands joined together and lifted up, as it were, deprecating something sitting in a four-footed Chair, which is carried by Four Winged Boys and they affirm that this Image increases felicity, riches, honors and confers benevolence and prosperity and frees from enemies.

They made, also, another Image of Jupiter, for a religious and glorious Life and advancement of fortune, whose figure was a Man having the Head of a Lion or a Ram and Eagle's Feet and clothed with Saffron colored Clothes.

Chapter 38
Of the Images of Mars

From the operation of Mars, they made an Image in the Hour of Mars (Mars ascending in the Second Face of Aries), in a Martial Stone, especially in a Diamond; the form of which was a Man armed, riding upon a Lion, having in his right Hand a naked Sword erect, carrying in his left Hand, the Head of a Man.

They report that an Image of this kind renders a Man powerful in Good and Evil, so that he shall be feared by all; and whoever carries it, they give him the Power of Enchantment, so that he shall terrify men by his looks when he is angry and stupefy them.

They made another Image of Mars, for obtaining boldness, courage and good fortune, in wars and contentions, the form of which was a Soldier, armed and crowned, girt with a Sword, carrying in his right Hand a long lance and they made this at the Hour of Mars, the First Face of Scorpio ascending with it.

(The Zodiac and The Mansions of the Moon)

Chapter 39
Of the Images of the Sun

From the operations of the Sun they made an Image at the Hour of the Sun, the First Face of Leo ascending with the Sun; the form of which, was a King crowned, sitting in a chair, having a Raven in his bosom and under his feet, a Globe. He is clothed in Saffron colored clothes. They say that this Image renders men invincible and honorable and helps to bring their business to a good end and to drive away vain dreams. Also, to be prevalent against fevers and the plague. And they made it in a Balanite Stone, or a Ruby, at the Hour of the Sun, when he, in his Exaltation, ascends Fortunately.

They made another Image, of the Sun in a Diamond, at the Hour of the Sun ascending in his Exaltation; the figure of which was a Woman Crowned, with the gesture of one, dancing and laughing, standing in a Chariot drawn by Four Horses, having in her right hand a Looking Glass or Buckler, in the left a Staff, leaning on her Breast, carrying a Flame of Fire on her Head. They say that this Image renders a man, fortunate and rich and beloved of all. And they made this Image on a Cornelian Stone at the Hour of the Sun ascending in the First Face of Leo, against Lunatic Passions which proceed from the combustion of the Moon.

Sun over a Martian Desert

Chapter 40
Of the Images of Venus

From the operations of Venus they made an Image, which was available for favor and benevolence at the very hour it ascended into Pisces, the form of which was the Image of a Woman, having the Head of a Bird, the Feet of an Eagle and holding a Dart in her Hand. They made another Image of Venus, to obtain the Love of Women, in the Lapis Lazuli, at the Hour of Venus, Venus ascending in Taurus, the figure of which was a Naked Maiden, with her Hair spread abroad, having a Looking Glass in her Hand and a Chain tied about her Neck; and near a handsome young man, holding her with his left hand by the Chain, but with his right hand doing up her Hair and both looking lovingly on at one another. And about them is a little Winged Boy, holding a Sword or Dart.

They made another Image of Venus, the First Face of Taurus, Libra or Pisces, ascending with Venus; the figure of which was a little Maiden with her Hair spread abroad, clothed in long White Garments, holding a Laurel Apple or Flowers, in her right Hand and in her left Hand, a Comb. It is said to make men pleasant, jocund, strong, cheerful and to give beauty.

Sun over a Venetian Desert

Chapter 41
Of the Images of Mercury

From the operations of Mercury, they made an Image of Mercury, Mercury ascending in Gemini, the form of which was a handsome young Man, bearded, having in his in left Hand a Rod, round which a Serpent was entwined; in the right he carried a Dart, having his Feet winged.

They say that this Image confers knowledge, eloquence, diligence in merchandise and gain; moreover, to obtain peace and concord and cure fevers.

They made another Image of Mercury, ascending in Virgo, for good will, wit and memory, the form of which, was a Man sitting upon a Chair, or riding upon a Peacock, having Eagle's Feet; and on his Head, a Crest and in his left Hand, holding a Cock of Fire.

Chapter 42
Of the Images of the Moon

From the operations of the Moon they made an Image for Travelers against weariness at the Hour of the Moon, the Moon ascending it its Exaltation; the figure of which was a Man leaning on a Staff, having a Bird on his Head and a Flourishing Tree before him.

They made another Image of the Moon for the increase of the Fruits of the Earth and against Poisons and Infirmities of Children, at the Hour of the Moon, it ascending in the First Face of Cancer; the figure of which was a Woman cornuted, riding on a Bull or a Dragon with Seven Heads or a Crab and she has in her right Hand a Dart and in her left Hand a Looking Glass, clothed with White or Green and having on her Head, Two Serpents with Horns twined together – and to each Arm, a Serpent is twined about – and to each Foot, a Serpent in like manner twined about.

And thus much is spoken here concerning the Figures of the Planets, that may suffice to our earliest instruction.

Chapter 43
Of the Images of the Head and Tail
Of the Dragon of the Moon

They made, also, the Image of the Head and Tail of the Dragon of the Moon, namely, between an aerial and fiery Circle, the likeness of a Serpent, with the Head of a Hawk, tied about them after the manner of the Great Letter Theta.

They made it when Jupiter, with the Head, obtained in mid Heaven, which Image they affirm to avail much for the success of petitions and would signify by this Image a good and fortunate genius, which they would represent by this Image of the Serpent.

For the Egyptians and Phoenicians do extol this Creature above all others and say it is a Divine Creature and has a Divine Nature. For in this is a more acute Spirit and a greater Fire than in any other, which thing is manifest both by his swift motion without feet, hands or any other instruments and also that it often renews its age with his skin and becomes young again. But they made the Image of the Tail like as when the Moon was eclipsed in the Tail or ill affected by Saturn or Mars; and they made it to introduce anguish, infirmity and misfortune. We call it an Evil Genius.

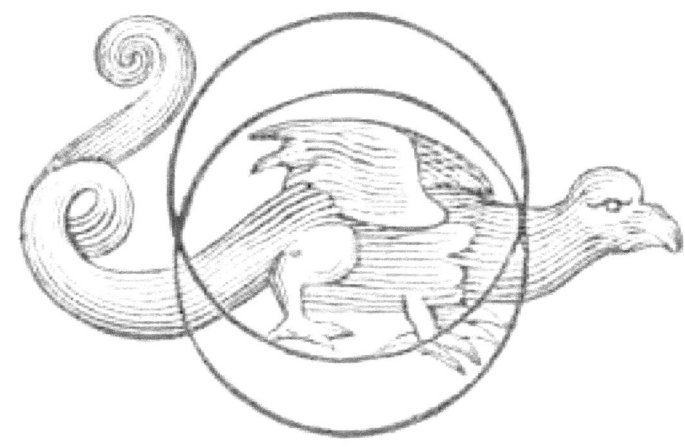

THE TALISMAN OF THE DRAGON'S HEAD

Chapter 44
Of the Images of the Mansions of the Moon

They made also, Images for every Mansion of the Moon as follows.

IN THE FIRST, for the destruction of someone, they made, in an Iron Ring, the Image of a Black Man, in a Garment of Hair and girdled round, casting a small Lance with his right hand. They sealed this in Black Wax and perfumed it with Liquid Storax and wished some Evil to come.

IN THE SECOND, against the Wrath of the Prince and for reconciliation with him, they sealed, in White Wax and Mastich, the Image of a King Crowned and perfumed it with Lignum Aloes.

IN THE THIRD, they made an Image of a Silver Ring, whose Table was Square; the Figure of which was a Woman, well-clothed, sitting in a Chair, her right hand being lifted upon her Head. They sealed it and perfumed it with Musk, Camphor and Calamus Aromatica. They affirmed that this gives Happy Fortune and every Good Thing.

IN THE FOURTH, for revenge, separation, enmity and ill-will, they sealed, in Red Wax, the Image of a Soldier sitting on a Horse, holding a Serpent in his right hand. They perfumed it with Red Myrrh and with Storax.

IN THE FIFTH, for the favor of Kings and Officers and good entertainment, they sealed, in Silver, the Head of a Man and perfumed it with Red Sanders.

IN THE SIXTH, to produce Love between Two, they sealed, in White Wax, Two Images, embracing one another and perfumed them with Lignum Aloes and with Amber.

IN THE SEVENTH, to obtain every Good Thing, they scaled, in Silver, the Image of a Man, well-clothed, holding up his Hands to Heaven, as it were, praying and supplicating; and perfumed it with Good Odors.

IN THE EIGHTH, for Victory in War, they made a Seal in Tin, being an Image of an Eagle, having the Face of a Man and perfumed it with Brimstone.

IN THE NINTH, to cause Infirmities, they made a Seal of Lead, being the Image of a Man, wanting his Privy Parts, covering his Eyes with his Hands; and they perfumed it with Rosin of the Pine.

IN THE TENTH, to facilitate child-bearing and to cure the Sick, they made a Seal of Gold, being the Head of a Lion and perfumed it with Amber.

IN THE ELEVENTH, for Fear, Reverence and Worship, they made a Seal of a Plate of Gold, being the Image of a Man riding on a Lion, holding the Ear thereof in his left Hand and in his right Hand, holding forth a Bracelet of Gold; and they perfumed it with Good Odors and with Saffron.

IN THE TWELFTH, for the Separation of Lovers, they made a Seal of Black Lead, being the Image of a Dragon Fighting with a Man; and they perfumed it with the Hairs of a Lion and Asafetida.

IN THE THIRTEENTH, for the Agreement of Married People and for dissolving of All the Charms against Copulation, they made a Seal of the Images of Both (of the Man in Red Wax and of the Woman in White Wax), and caused them to Embrace one another; perfuming it with Lignum Aloes and Amber.

IN THE FOURTEENTH, for Divorce and Separation of the Man from the Woman, they made a Seal of Red Copper, being the Image of a Dog Biting his Tail; and they perfumed it with the Hair of a Black Dog and a Black Cat.

IN THE FIFTEENTH, to obtain Friendship and Goodwill, they made the Image of a Man Sitting and Indicting Letters; and perfumed it with Frankincense and with Nutmegs.

IN THE SIXTEENTH, for Gaining Much Merchandising, they made a Seal of Silver, being the Image of a Man, Sitting on a Chair, Holding a Balance in his Hand; and they perfumed it with Well-Smelling Spices.

IN THE SEVENTEENTH, against Thieves and Robbers, they Sealed with an Iron Seal, the Image of an Ape; and perfumed it with the Air of an Ape; (the Essence or Sweat and Saliva of the Ape).

IN THE EIGHTEENTH, against Fevers and Pains of the Belly, they made a Seal of Copper, being the Image of a Snake with his Tail above his Head; and they perfumed it with Hart's Horn; and said this same Seal put to flight Serpents and All Venomous Creatures from the Place where it is Buried.

IN THE NINETEENTH, for facilitating Birth and provoking the Menstrues, they made a Seal of Copper, being the Image of a Woman holding her Hands upon her Face; and they perfumed it with Liquid Storax.

IN THE TWENTIETH, for Hunting, they made a Seal of Tin, being the Image of Sagittarius, Half a Man and Half a Horse; and they perfumed it with the Head of a Wolf.

(Concerning the Twentieth Image of the Mansions of the Moon. Referring the instruction, "perfumed it with the Head of a Wolf". Possibly explicably something from the origin of his mouth, but in the case of this text, it cannot be excluded that he will have the meaning of the actual brains of the beast, whatever he refers to. The specificity of instruction in all the terms of the text are not as important as the conveyance of the text by itself.)

IN THE TWENTY-FIRST, for the Destruction of Somebody, they made the Image of a Man, with a Double Countenance Before and Behind; and they perfumed it with Brimstone and with Jet and put it in a Box of Brass and with it, Brimstone and Jet and the Hair of Him whom they would Hurt.

IN THE TWENTY-SECOND, for the Security of Runaways, they made a Seal of Iron, being the Image of a Man, with Wings on his Feet, bearing a Helmet on his Head; and they perfumed it with Argent Vive.

IN THE TWENTY-THIRD, for Destruction and Wasting, they made a Seal of Iron, being the Image of a Cat, having a Dog's Head; and they perfumed it with Dog's Hair taken from the Head and buried it in the Place where they intended the Hurt.

IN THE TWENTY-FOURTH, for multiplying Herds of Cattle, they took the Horn of a Ram, or Goat or that sort of Cattle they would increase; and sealed in it, burning, with an Iron Seal, the Image of a Woman giving her Son his milk; and they hung it upon the Neck of that Cattle who was the Leader of the Flock, or they sealed it in his Horn.

IN THE TWENTY-FIFTH, for the preservation of Trees and Harvest, they sealed in the Wood of a Fig Tree, the Image of a Man planting and they perfumed it with the Flowers of the Fig Tree and hung it on the Tree.

IN THE TWENTY-SIXTH, for Love and favor, they sealed in White Wax and Mastich, the Figure of a Woman washing and combing her Hair and they perfumed it with Good Odors.

IN THE TWENTY-SEVENTH, to destroy fountains, pits, medicinal waters and baths, they made of Red Earth, the Image of a Man Winged, holding in his Hand, an Empty Vessel that was broken. The Image being burnt, they put into this Vessel, Asafetida and Liquid Storax. And they buried it in the Pond or Fountain which they would Destroy.

IN THE TWENTY-EIGHTH, for getting Fish together, they made a Seal of Copper, being the Image of a Fish. And they perfumed it with the Skin of a Sea Fish and cast it into the Water where they would have the Fish gathered.

Moreover, together with the aforesaid Images, they wrote down also the Names of the Spirits and their Characters and invoked and prayed for those things which they pretended to obtain.

(Asafetida or Mugwort, nicknamed here, Chicory Sorrelspore)

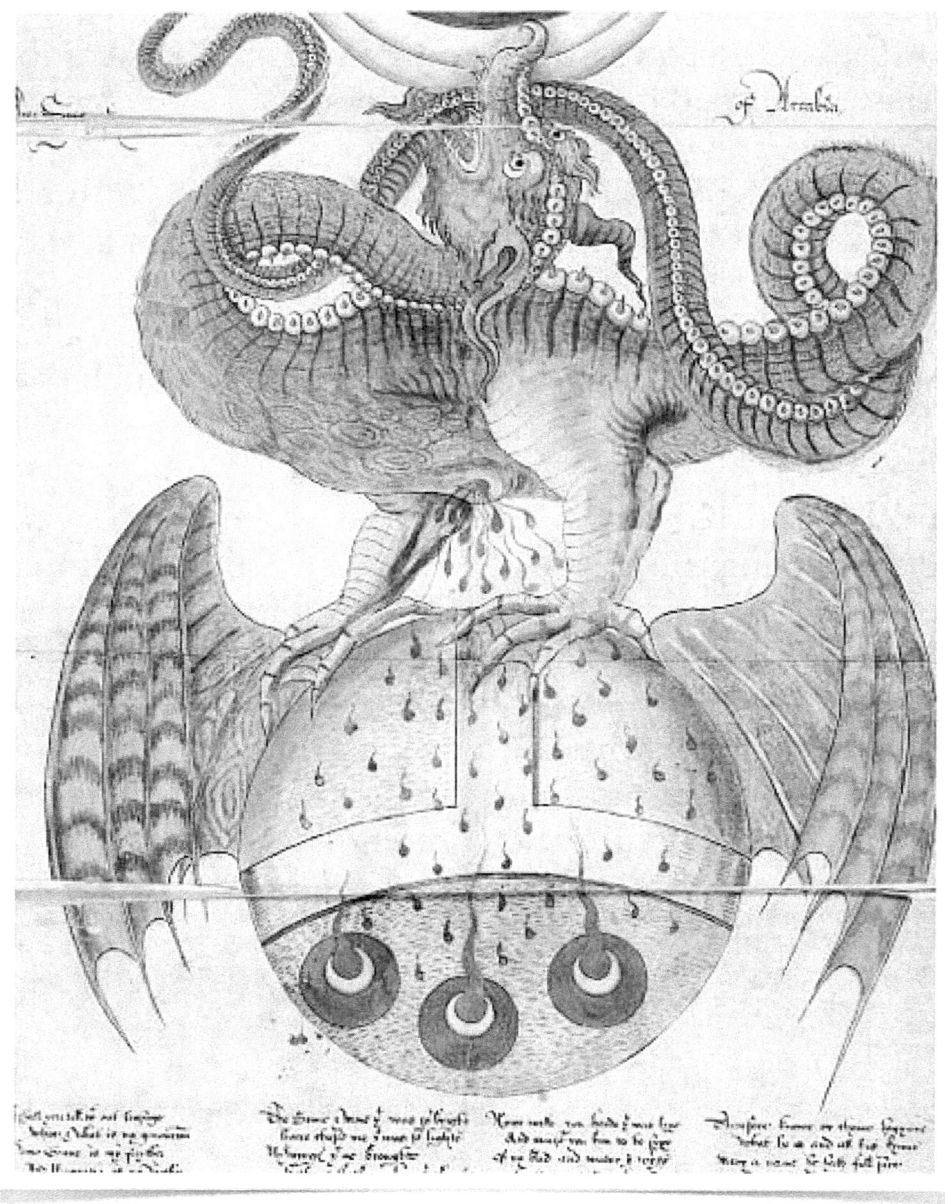

The Dragon is the Master of the Universe and the interlocutor beyond angels or other ephemeral creatures; the orator and articulator of all mystery besides Christ; the mythological unknown of all the realms, between those physical and those astral. We can never know for instance through this treatise whatever is the magical truth of the place of mythological creatures in the celestial and super-celestial, because he refuses to see any magical creatures in the natural order; but that they are identities and later to be called all angels.

Chapter 45
That Human Impressions
Naturally Impress their Powers upon External Things
and How Man's Mind, through a Degree of Dependencies,
Ascends into the Intelligible World
and becomes like to the More Sublime Spirits and Intelligences

The Celestial Souls send forth their Virtues to the Celestial Bodies, which transmit them to this Sensible World; for the Virtues of the Terrene Orb (Terrestrial Globe) proceed from no other Cause than Celestial.

Hence the Magician, that will work by them, uses a cunning Invocation of the Superiors, with Mysterious Words and a Certain Kind of Ingenious Speech, drawing the one to the other; yet (he does his provocative works) by a Natural Force, through a certain mutual agreement between them, whereby things follow of their own accord, or sometimes are drawn unwillingly.

Hence says Aristotle, in this Sixth Book of his Mystical Philosophy, that, "When anyone, by binding or bewitching, calls upon the Sun or other Stars, praying them to assist the work desired, the Sun and other Stars do not hear his words, but are moved, after a certain manner, by a certain conjunction and mutual series, whereby the Parts of the World are mutually subordinate the one to the other and have a mutual consent, by reason of their Great Union.

"As in a Man's Body, one Member is moved by perceiving the motion of another; and as in a Harp, One String is moved by the motion of another. So when anyone moves any Part of the World, other Parts are moved by the perceiving of that motion."

The Knowledge, therefore, of the Dependency of Things following one to the other, is the Foundation of all wonderful operation, which is necessarily required to the exercising the Power of attracting Superior Virtues. Now the Words of Men are certain Natural Things. (Words serving for some of the conceptual means of the intellect to attain the realities of all these wonderful operations.)

And because the Parts of the World mutually draw one to the other; therefore, a Magician, invocating by words, works by Powers fitted to Nature, by leading some by the love of one to the other, or drawing others, by reason of the one following after the other, or by repelling, by reason of the enmity of one to the other, from the contrariety and difference of things and the multitude of Virtues.

Which (contrary things), although they are so different, yet they are Perfect to one Part. Sometimes, also, he compels things by way of authority, by the Celestial Virtue, because the Magician is not a Stranger to the Heavens. A Man, therefore, if he receives the Impression of a Ligation, or Fascination, does not receive it according to the Rational Soul, but Sensual; and if he suffers in any Part, he suffers according to the Animal Part. For they cannot draw a Knowing and Intelligent Man by Reason, but by receiving that impression and force by Sense; in as much as the Animal Spirit of Man is, by the influence of the Celestials and co-operation of the Things of the World, affected beyond his former and natural disposition.

As the Son moves the Father to labor, although unwilling, the Father will keep and maintain his Son, although he be wearied; and even so, the desire to rule is moved by anger and other labors, to get to the point of dominion; and as the indigence of Nature and the fear of Poverty, moves a man, it moves him to desire riches; and as the ornaments and beauty of a Woman is an enticement to concupiscence, so is the way with them; and as the harmony of a Wise Musician moves his hearers with various Passions, some listeners do voluntary follow the consonance of art, others will conform themselves to enjoyment by gesture, although unwilling, because their sense is captivated and their reason not being intent to these natural things.

Hence they fall into Errors, who think those things to be above Nature or contrary to Nature – which indeed are by Nature and according to Nature. We must know, therefore, that every superior moves its next inferior, in its degree and order, not only in Bodies, but also in Spirits; so the Universal Soul moves the particular Soul. (Nature is the prime mover of all lesser bodies in natural order.)

The Rational acts upon the Sensual and that upon the Vegetable and every part of World acts upon another and every part is apt to be moved by another. And every part of this Inferior World suffers from the Heavens, according to their Nature and aptitude, as one part of the Animal Body suffers for another. And the Superior Intellectual World moves all things below itself; and, after a manner, contains all the same Beings, from the First to the Last, which are in the Inferior World.

Celestial Bodies, therefore, move the Bodies of the Elementary World, compounded, generable, sensible (from the circumference to the center), by Superior, Perpetual and Spiritual Essences, depending on the Primary Intellect, which is the Acting Intellect; but upon the Virtue put in by the Word of God; which Word, the Wise Chaldeans of Babylon call, the Cause of Causes; because from it are produced all Beings.

The Acting Intellect, which is the Second, from it depends; and that by reason of the Union of the Word with the First Author, from Whom, All Things Being, are truly produced.

The Word, therefore, is the Image of God ~ the Acting Intellect, the Image of the Word. The Soul is the Image of this Intellect and our Word is the Image of the Soul, by which it acts upon Natural Things naturally, because Nature is the work thereof.

(The entire schema explained in this page alone explains the method of intellectual obtaining in the Magicians' power.)

And every one of those perfects his subsequent: as a father his son; and none of the latter exists without the former; for they are depending among themselves by a kind of ordinate dependency – so that when the latter is corrupted, it is returned into that which was next before it, until it come to the Heavens; then to the Universal Soul; and, lastly, into the Acting Intellect, by which all other creatures exist; and itself exists in the Principal Author, which is the creating Word of God, to which, at length, All Things are returned. Our Soul, therefore, if it will work any wonderful thing in these inferiors, must have respect to their beginning, that it may be strengthened and illustrated by that and receive power of acting through each degree, from the very First Author.

Therefore we must be more diligent in contemplating the Souls of the Stars – (more so) than their Bodies; and (also more diligent to contemplate) the Super-Celestial and Intellectual World – (more so) than the Celestial (or) Corporeal, because that is (a more noble thing to do, a better exercise of faith and science). Although, the Celestial and Corporeal path also is an excellent way to that other, higher path of action and knowledge, and without which mediumship in the first case, the influence of the superior cannot be attained.

As for example: the Sun is the King of Stars, most full of Light; but receives its Light from the Intelligible World, above all other Stars, because the Soul thereof, is more capable of Intelligible Splendor.

(Therefore, we must understand that the body of a Soul has the intention in the attainment of its spiritual goals, to obtain the higher intellect of the Super Spiritual, or the Super Celestial, otherwise, the Astral Plane; and therefore, also, through this scientific approach of contemplation and meditation in the truth of cosmological things, to see the body evolving into the astral environ and its existence. This would be weather a body is celestial or whether a body has an animal or human form; the soul in the world, proceeds in intellect, from the body and the attainment must be a joining together of the goal for the higher plane and a determination of its location in the incorporeal, or astral, otherwise, intellectual (or mental) super-celestial.)

Wherefore he that desires to attract the influence of the Sun, must contemplate upon the Sun; not only by the speculation of the exterior light, but also of the interior.

And no man can do this, unless he return to the Soul of the Sun and become like to it and comprehend the Intelligible Light thereof with an Intelligent Sight, as the Sensible Light with the Corporeal Eye. For this Man shall be filled with the Light thereof and the Light whereof; which is an under-type impressed by the Supernal Orb, it receives into itself, with the Illustration whereof, his Intellect, being endowed and truly like to it and being assisted by it, shall at length attain to that Supreme Brightness and to all forms that partake thereof. And when he has received the Light of the Supreme Degree, then his Soul shall come to Perfection and be made like to Spirits of the Sun and shall attain to the Virtues and Illustrations of the Supernatural Virtue and shall enjoy the Power of them, if he has obtained Faith in the First Author.

In the first place, therefore, we must implore assistance from the First Author and, praying, not only with mouth, but a religious gesture and supplicant Soul, also abundantly, incessantly and sincerely, that he would enlighten our Mind and remove darkness, growing upon our Souls by reason of our Bodies.

(We should pray to the Lord of all Creation, that we could have our physical lives enriched and fulfilled to an understanding of the completion of the universe as it has us in an organized place in it.)

(This last promise in Chapter 45 of the Sublime Intelligible Realm of the Sun and the remission of Man's Physical Form into the Universal Light, so interpreted, is an Ancient Cosmogony as expressed by the Egyptian Worship of the Sun God, Ra. So we can believe that much or most, if not all, of this mysterious doctrine presented by Francis Barrett, thus far, is based and rooted in the Ancient Cosmogony of the Egyptians and not "made up"; or so I conclude.)

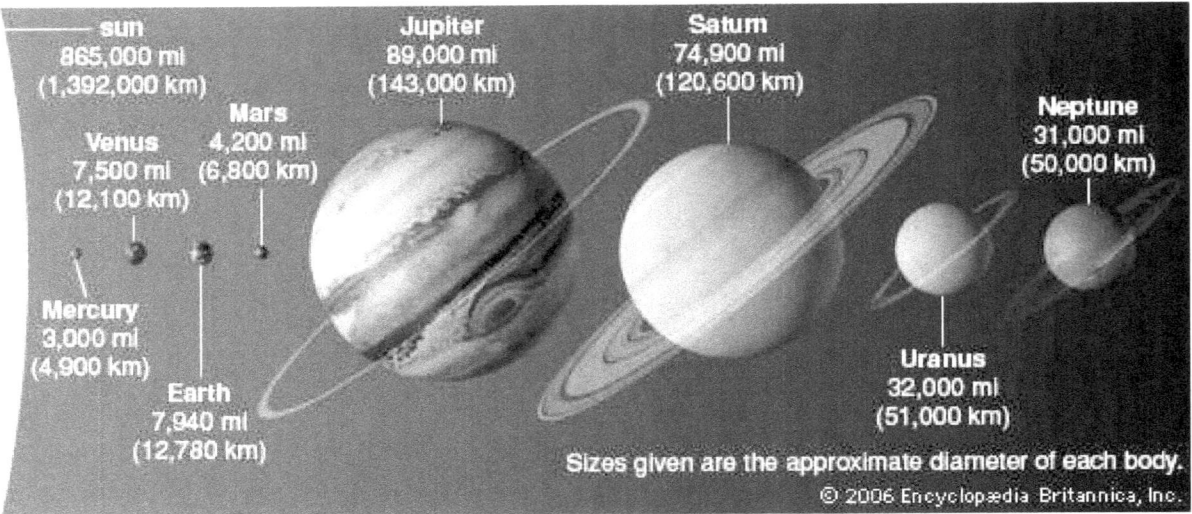

(Somewhere between conception and realization. A picture becomes alive to its own meaning to become reality in the mind of its keeper.)

(But beyond doubt, a better intellectual understanding of the astral universe and its co-linear time relation to the material and planar universe competing for it, would allow us to dream more largely with confidence over learning the meaning of our hopes.)

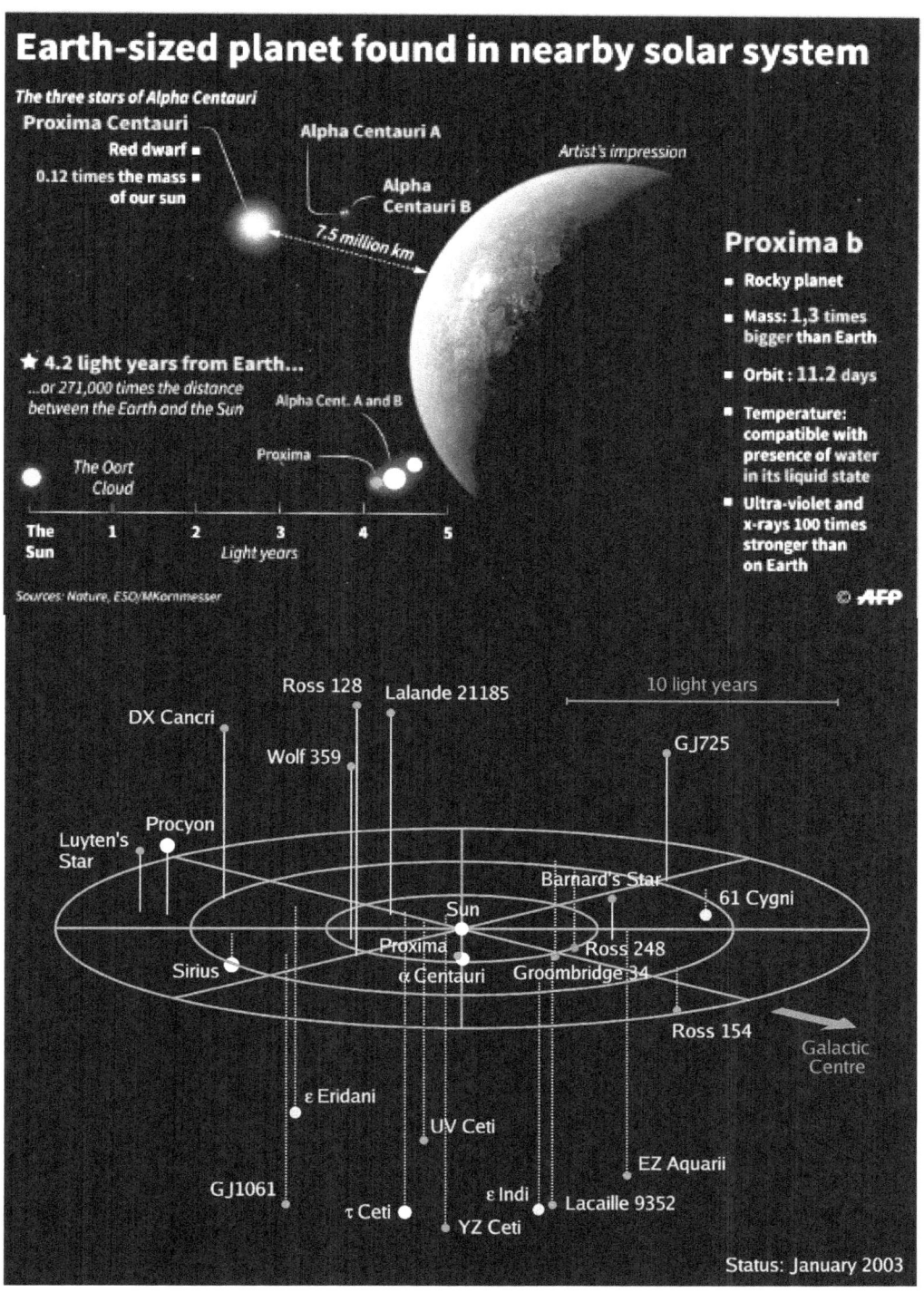

Chapter 46
The Conclusion of the Constellatory Practice,
or, Talismanic Magic;
in which is included ~
the Key of All that has been written upon this subject;
showing the Practice of Images, etcetera,
by way of example, and likewise ~
the necessary Observations of the Celestials,
towards the Perfection of Talismanic Operations

We will now show you the Observations of Celestial Bodies, which are required for the practice of these things, which are briefly as follows.

To make anyone Fortunate, we make an Image at that time in which the Signification of Life, the Giver of Life, or Hylech, the Signs and Planets, are Fortunate.

Let the Ascendant and mid-Heaven and the Lords thereof, be Fortunate.

And also, the Place of the Sun and Moon, being part of Fortune and the Lord of conjunction or prevention, make (this) before their nativity, by depressing the Malignant Planets, that is, taking the times when they are depressed.

But if we would make an Image to procure Misery, we must do contrary to this.

And those which we before placed Fortunate, we must now make Unfortunate, by taking the Malignant Stars when they rule. And the same means we must take to make any place, region, city or house Unfortunate.

But if you would make anyone Unfortunate who has injured you, let there be an Image made under the Ascension of that man whom you would make Unfortunate.

And you shall take, when Unfortunate, the Lord of the House of his Life, the Lord of the Ascendant and the Moon, the Lord of the House of the Moon, the Lord of the House of the Lord Ascending and the Tenth House and the Lord thereof.

Now, for the building, success or fitting of any place, place Fortunes in the Ascendant thereof and in the First and Tenth, ~ the Second and Eighth House, you shall make the Lord of the Ascendant and the Lord of House of the Moon, Fortunate.

But to chase away certain Animals (from any place) that are noxious to you, that they may not generate or abide there, make an Image under the Ascension of that Animal which you would chase away or destroy and after the likeness thereof.

For instance, now suppose you would wish to chase away Scorpions from any place. Let an Image of a Scorpion be made, the Sign Scorpion ascending with the Moon.

Then you shall make Unfortunate the Ascendant and the Lord thereof and the Lord of the House of Mars. And you shall make Unfortunate, the Lord of the Ascendant in the Eighth House. And let them be joined with an aspect Malignant, as opposite or square. And write upon the Image, the Name of the Ascendant and of the Lord thereof and the Moon, the Lord of the Day and Hour.

And let there be a Pit made in the Middle of the Place from which you would drive them out and put into it some Earth taken out of the Four Corners of the same place, then bury the Image there, with the Head downwards, saying, "This is the burying of the Scorpions, that they may be forced to leave and come no more into this place."

And so do by the rest.

Now for gain, make an Image under the Ascendant of that man to whom you would appoint the gain. And you shall make the Lord of the Second House, which is the House of Substance, to be joined with the Lord of the Ascendant, in a trine or sextile aspect and let there be a reception among them.

You shall make Fortunate the Eleventh and the Lord thereof and the Eighth House. And, if you can, put part of Fortune in the Ascendant or Second House. And let the Image be buried in that place or from that place, to which you would appoint the gain or Fortune.

Likewise, for agreement or love, let be made an Image in the Day of Jupiter, under the Ascendant of the Nativity of him whom you would wish to be beloved. Make Fortunate the Ascendant and the Tenth House and hide the Evil from the Ascendant.

And you must have the Lords of the Tenth and Planets of the Eleventh, Fortunate, joined to the Lord of the Ascendant, from the trine or sextile, with reception.

Then proceed to make another Image, for him whom you would stir up to love, whether it be a friend or female or brother or relation or companion of him whom you would have favored or beloved; if so, make an Image under the Ascension of the Eleventh House from the Ascendant of the First Image.

But if the party be a wife or husband, let it be made under the Ascension of the Third House; if the person is a mother, let the Ascension be of the Tenth House and so on.

Now let the signification of the Ascendant of the Second Image be joined to the signification of the Ascendant of the First Image and let there be between them a reception and let the rest be Fortunate, as in the First Image.

Afterwards join both the Images together in a mutual embrace or put the Face of the Second Image to the back of the first and let them be wrapped up in silk and cast away or spoiled.

Also, for the success of petitions and obtaining of a thing denied or taken or possessed by another, make an Image under the Ascendant of him who petitions for the thing and cause the Lord of the Second House to be joined with the Lord of the Ascendant, from a trine or sextile aspect and let there be a reception between them.

And, if it can be so, let the Lord of the Second House be in the obedient signs and the Lord of the Ascendant be in the ruling; and make Fortunate the Ascendant and the Lord thereof.

And beware that the Lord of the Ascendant be not retrograde or combust or cadent or in the House of Opposition; that is, in the Seventh House from his own House.

Let him not be hindered by the Malignant Planets, but let him be strong and in an angle. You shall make Fortunate the Ascendant and the Lord of the Second House and the Moon.

And make another Image for him that is petitioned to and begin it under the Ascendant belonging to him, as if he is a King or Prince or Noble and begin it under the Ascendant of the Tenth House from the Ascendant of the First Image; if a father, under the Fourth House and if a son, under the Fifth House and so of the like.

Then put the signification of the Second Image, joined with the Lord of the Ascendant of the First Image from a trine or sextile and let him receive it. And put them both strong and Fortunate, without any hindrance. Make all evil from them.

You shall make Fortunate the Tenth and the Fourth Houses, if you can, or any of them. And when the Second Image shall be perfect, join it with the First, face to face and wrap them in clean linen and bury them in the middle of his house who is the petitioner, under a Fortunate Signification, the Fortune being Strong.

And let the Face of the First Image be towards the North, or rather towards that place where the thing petitioned does remain; or, if it happens that the petitioner goes forward to obtain the thing desired or petitioned for, let him carry the said Images with him.

Thus we have given, in a few examples, the Key of All Talisman Operations whatsoever, by which wonderful effects may be wrought either by Images, by Rings, by Glasses, by Seals, by Tables, or any other Magical Instruments whatsoever.

But as these have their chief grounds in the True Knowledge of the Effects of the Planets and the Rising of the Constellations, we recommend an earnest attention to that part of Astrology which teaches of the Power, Influences and Effects of the Celestial Bodies among themselves generally.

Likewise, we would recommend the Artist to be Expert in the aspects, motions, declinations, risings, etcetera, of the Seven Planets and perfectly to understand their Natures, either Mixed or Simple; also, to be ready and correct in the erecting of a Figure at any time, to show the true Position of the Heavens, there being so great a sympathy between the Celestials and ourselves; and to observe all the other Rules which we have plentifully recited.

And, without doubt, the industrious student shall receive the satisfaction of brining his operations and experiments to effect that which he ardently desires.

With which, wishing all success to the Contemplator of the Creature and the Creator, we will here close up this Second Part of our Work and the Conclusion of Our Book of Talisman Magic.

This Ends the First Book of The Magus by Francis Barrett.

The Second Book appears in a separate Volume.

A Type of Speculated Venue for the Future of Mankind, though Unlikely

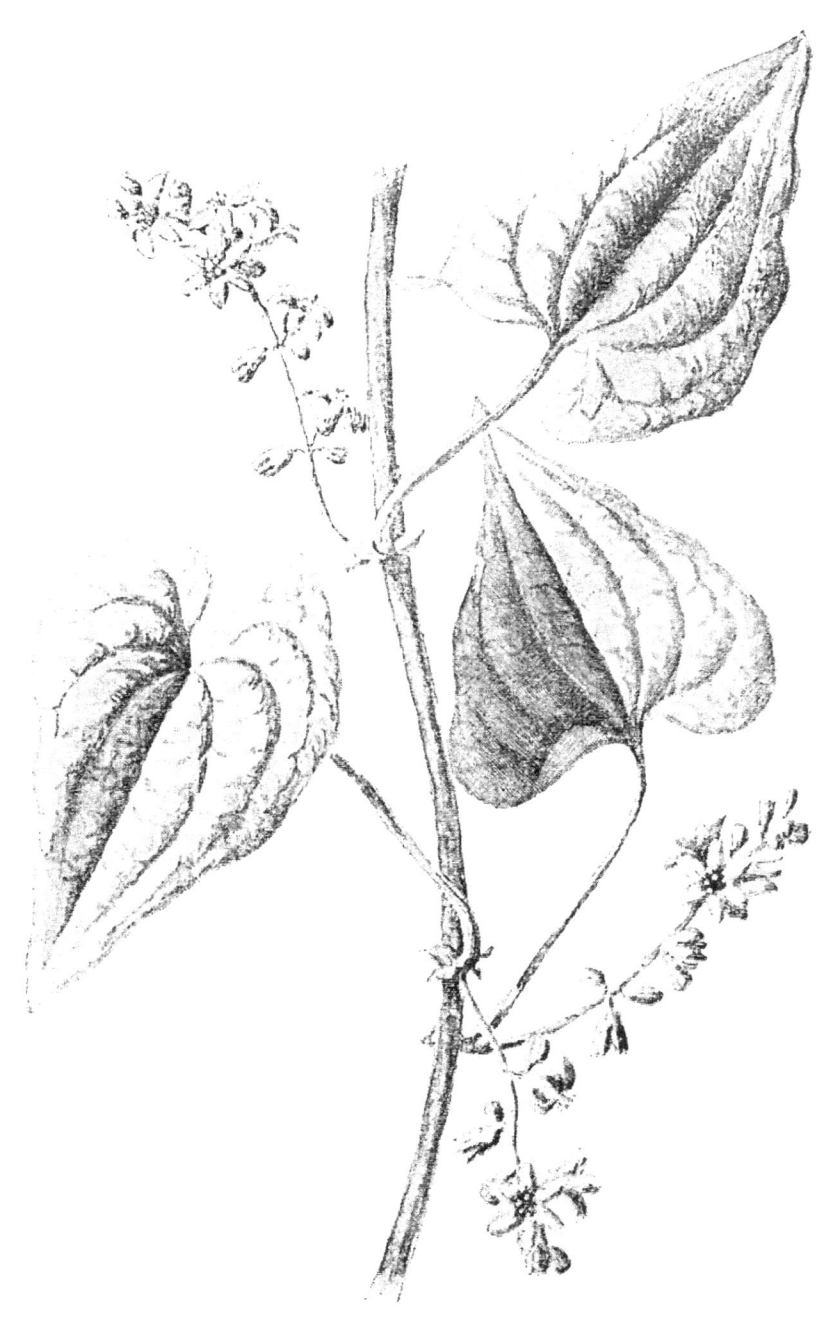

Eliphas Levi's Tetragrammaton Pentagram, which he considered to be a Symbol of the Microcosm, or, Human Being – (Wikipedia Reference and Image)

Ending Notes ~ by Patricia Spencer ~ for the Complete Text

The above diagram of the Pentagram is simply given to show it.

It was the best representation of the symbol found to begin the discussion of such a working symbol of the Pentagram. It seems to bring itself to the point then and therefore of a usable Talisman under the cosmological domain of many philosophies, schools of thought and faiths, besides only that of modern esoteric practices in the Occult; but for the fact that it may also contain a relevant amount of "power", if that is what the treatise means for magic, over the astrophysical to physical element, associating itself to faith, philosophy and thought by the term "metaphysical"; and as such, "metaphysical" term, also in the connotation of being a "co-" group of two astrophysicals, which "there" otherwise to conceive the altercation of "meta" is celestial to physical, and which here is in the ordinary of the natural and physical; so it is I have added metaphysical as an important term for the establishment of the symbol as a Talisman for magical principles.

In such case of judging it, the Pentagram Talisman may be considered a valuable Talisman for the "power" of a kind of "magic" in arguable theories of metaphysics and the legalistic possibilities thereof, for magic to be established as a school of thought and practice within its domain.

And what I mean by legalistic, is through the exercise of some given course of reason, which is legalistic in its value, to have concurrence with creditable facts in the course of philosophy (as stated to itself), rather than in the occult of lost sciences that will differ from the knowable methods of philosophical inquiry.

That is to say, by the term of establishing facts to mind, which are both theoretical and independent of physical causes in the first case and otherwise empirical only in the second. And whereas, the secondary rationale is not as important as the primary to determining the truth of the empirical values.

Whereas, the school that is magic itself does not have the theoretical demand upon itself that philosophy demands of all students in order to obtain a rationale of aesthetic reasons influencing the outcome of judgment primarily and preceding the rationale of physical reasons alone. Magic per se as known, has no greater distinction than to seem physical and to rule empirical philosophy; however, many philosophers who have a use for magic in their literatures, have also a requirement (as Francis Barrett does most definitively, though his philosophy is theological for the most part and not merely in the aesthetic range of schemata, as others might demand); I begin again, many philosophers have also an excellent added requirement for the proof of magic, that it demands for the proof of itself besides a physical outcome but in a theoretical outcome without of the domain of singularly crucifying matter to find its esoteric material value, or essence and to use the product of it in the plethora or Pleroma of the unknown outer to inner- space domain of application. So compiling all these requirements together of magic, empiricism, philosophy, aesthetics, reason, so on, we find ourselves in the domain together with everyone consenting, that is named ~ cosmogony; and that is therefore, the occult science of which these prophets of the discipline speak; prophets as in of the rationalism and future empirical of the teaching.

So this work, this book by Barrett in the first part, does generally demand a reform of magic into the domain of occult science, as cosmogony and into the aesthetic domain of ephemeral theory, in order to reform human ideals of the esoteric topic in philosophy.

However, as for the issue that most of cosmogony is lost to us as practicable, as we truly must see in the usage even of the astrological height to which we must correspond in order to practice the smallest of powers, there should still be sought out the modern equivalencies and relics and archaeologies which could indicate the path to the revelations of whatever is scientific from the lost articles of cosmogony as an Ancient System.

Since I am at a loss for how this might be done, I have given my perhaps both feeblest and also maybe most controversial suggestion in this Talisman; that, if even in the idea of separating it from the modern occult with witchcraft (and even if it were to be for the Satanic, this still accomplishes a separation from the ideal, as stated in the beginning notes, that Satan and the Devil are "Ko", as some will disagree, as do I; it would result in better judgments and values on cosmogonies as less destructive to natural order than any destructive ideals to nature usually are in that respect).

I've lost myself. But to say, to avoid the constant protection which this Idol Talisman (I say Idol because it is such of the both Ancient Testament of Our Scriptures and Legends of Faith, such as the Holy Grail in the newer Pseudepigraphia) for the Satanic Folklore of its present, common use, it might be considered an equitable Talisman for the exercise of metaphysical investigation indulging of the ancient cosmogonies and their magical declarations lost to us.

There must be more like this, but this is presently my favorite.

Not realizing if these words might be of help to anyone reading this text to understand the superficial wit I give to its topical presentation, I thought I might add the Wikipedia entry for the Talisman in its very diversified history leading up to the present types and times of application for it.

But this is then also the time to state here finally, as I wanted to try to be as silent of a gloss as I might be in the upper text, except where I thought I might be of some attribute in my undertaking an statement – that; although I do not believe this is a superficial text in any means, it is quite a beginner's course in a multitude of means for the believer to leave off the term, "esoteric," and to begin to at least consider cosmogony as belonging to a type of science which includes philosophy, with also the personal responsibility (on the part of the reader), to learn and to learn also to declare a system of faith over whatever magic will be learned.

All this, since it is true that, the universality of this science, seems to like to take care of itself as believable and plausible, without our opinion of its self-believing fraud to our natural experiences of cognizance; does not this alone indicate the universal cosmogony is a scientific domain independent of our recognition and therefore of itself, does indicate a separate science which holds an existence ephemeral of rule that is intrinsic of its own self-knowledge. But people (and even beasts) will believe that there is nothing knowable or scientific of a separate nature from our own creation.

But cosmogony does not require our approval so it seems and I say here, in that action alone, we could believe that there exists a science ruling; and as a necessity of the possibility for universal faith in general, we also understand that we are left outside of this pertaining knowledge most definitely: without Messianic leadership and without any human understanding. But the rule of cosmogony calls its demands for fulfillment within the self-intrinsic universal terms of nature, so that it seems to us, necessarily primordial and basic and also wild and proficient of a natural means inherent in both body and mind of the creature status of both humanity and bestiality. And so we realize the connection to the occult in natural science constantly but we have no understanding of its approach. Therefore, the long treatise of its unknowable domain by many authors and my own ending entreaty to ponder the ideology of so much responsibility.

So human beings, if that's what we are as separated from beasts, must look to the approach of the topic of magic, (that being outside of the domain of science and naturalism, outside of the domain of astrophysical and theoretical cosmogony and rationalism, outside of bestial natural astrophysical order), by finding their own reason at every step of the approach. This is the grand point, which, falling from thematic and episteme application and recognition of, will lend itself back to huge sorrows of myriad and diverse kinds. What a trifling point to overlook, when the marriage of Mankind and beasts together may be united again with the prime and perfect order thereby and suspectly. But we say no, being responsible, and lend the admonishment to everyone in faith of universal fellowship.

Let's discuss our point of view as human beings, separated from beasts again; and also hope to extend our responsibilities of universal fellowship to natural beasts thereby as well.

Persons are required of their very creature status (as separated from the bestial instinct), to the exercise of their reason and also faith where found out; which duo, nevertheless must be discriminating to its own accountability of talents and persuasions of which there is such diversity, there is scarce a friend of repute or report to consult; therefore, more reasons for methods of principled order. And as for our bestial counterparts, should we not hope to underestimate their experiences in the same cosmological universe as determinant of their own instincts and talents? Since, otherwise, our worse assumptions on them, might surely leave us void of their place in the temporal present and ephemeral hereafter; why should we leave them behind in our concerns, even though we have the added curse of keeping ourselves distinct and separate; wondering always, (as I do anyway), if they believe we are some other beasts or they are some other humans?

And but, speaking of ourselves, we must hopefully come to some kind of scientific conclusions and obviously, for the case of these higher ideals being difficult to seem fulfilled in the anatomy of science; for anyone, that is; to also conclude and declare the science of the species of beliefs and their resultant powers, with some kind of credible statement of philosophy which comprises faith together with reason. As to the subject of himself as the inquisitor of the esoteric realm, do we not understand that most of our energies are, as civilized, already exploited in this endeavor for the questions of plausible universalism without magic? So how much more difficult could it be to extend the same intellectual energy again for the purpose of extending faith and reason to include the absolutely implausible, such as a system of scientific cosmogony with astrophysical and religious implications to add to it. We confirm a determinism to be responsible.

But the primary point I am trying to prove and I hope I have put some conviction to it for the reader overall, concerns the exercise of metaphysical inquiry and its method in application to magic; that it is a necessary assumption for the establishment of principles.

Whereas, we might say with some security contrarily, that magic without principles, has its severest outcomes for any practitioner. So a mental, theoretical method is a requisite method and this is considerably one which includes faith with reason for the difficulty of its work being cosmogony, or the basic unknown and prime of the unknown, and which explains itself finally as no less than all those descriptors at one time, to mean to us, a metaphysical system.

So it is for that reason, that I have presented the text, since, we see in other disciplines of metaphysics, that the topics of both magic and cosmogony comes heavily under fire and it has some power to destroy the order of both faith and reason along with the physical universe and its implication for the outer universe at large.

Magic should be everyone's topic at some point and not just in childhood. But we have no respect for the undertaking and early abandon it. So moving on finally, we should have some presentable matter to study to begin in and as a course of its own (cosmological) advancement and not as the course of another discipline to begin; otherwise, we continue to avoid the topic.

And it is hopeful, that it simple enough at the beginning at least, to continue to trust that the Lord of all Creation, who does in fact know more than we could comprehend; and do we comprehend that and continue to strive to see its truth continually in our study. If we do not, then, adapt some principles of order in the study of magic, we might give ourselves some credible manner of self-regard that is higher than what we know and therefore, again, lose all of everything gained from being silent and studying as would be the case of any discipline at outset to determine it. So doing, and setting aside some time to read a few thoughts on the topic by some priest of the craft, with the proper precautions noted to our limited physical faith that should seem obvious to us by this time, we are generally set to the task of becoming simple students of the occult and remaining true to our own self-determinism as given at the origin.

So this book by Francis Barrett is only a beginning. The presentation of it should be understood without the necessary faith to practice all of its idols to fruition of proof; but to undertake the subject and topic and to more largely than anything else, have the lessons in mind credibly to the greater case of need for growing faith in a garden of deceit and conceits. That great case of need, being, that we have a responsibility to ourselves that is already difficult as it is to believe in our own knowledge at origin. And thus, to continue to support ourselves is to continue to fall into the faithless determinations of having learned nothing; whereas, if we see that the question is that in itself, we understand that we have to look, like examiners, into the possibilities wherever life may give itself chance. Because the greatest controversy of magic is the reversal of all living into death and its ritual evil, as we all understand the controversial implications. And to refuse the extended and open arm of science as we understand it in favor of the occult is an irresponsible way of life, but yet, in itself, it continues to budge its way into all of our necessary effects with science. At some point, we have to realize its existence.

I am merely suggesting a stratagem.

In any event of these admonitions and aspirations to be convincing with words, particularly those of metaphysic and cosmogony together; I suppose I haven't said much to justify what remains impractical and un-practicable of this primary text at offer; except in the admonishments themselves and also, except as the practice of a theoretical discipline, an intercession and integration, moreover, between these particular disciplines (metaphysics, cosmogony); then we very much realize so much of practical and ancient relevance and possibility.

But there are other texts and many as is well known. My choice was with Francis Barrett because of his relation to all the most primary topics and his determinism in philosophy.

Sidelight to this presentation, I hope that I will be able to complete the Second Book of this text in due season, or I mean, sometime, to seem relevant to those who will read this text soon.

But the next section of it deals succinctly with the study of the Cabal and Cabalists and not so much with Natural and Occult Philosophy together here as shown.

Part Two of the Magus by Francis Barret, regards the actual ascription of Identities to these Cosmogonies and the laws of these Identities according to a system of higher metaphysics, which is sometimes referred to as aesthetics, which remains to be seen to the serious student of magic, as an even more nepharious area of philosophical research than metaphysics and cosmogony, our two partners of the agreed debate, than there might be angels to end the fall of all creation by cataclysms of debates, since aesthetics in philosophy, takes up the ephemeral as beyond the celestial and meaning it as also accounting for the "real" and that meaning, the empirical. Barrett and practitioners of learned magic refer to this aesthetic systematic law as Cabalism, referring to ephemeral angels and their identities as the very principals and laws of science.

I hope before entering on to the presentation of Cabala writ, in the text for Part Two (which will not appear anywhere herein), to continue on a little bit longer with the presentation of some other author's more extempore study on the Occult and Natural Philosophy. I am interested in providing some factual basis to the contrary heated debate of the Occult as deadly to both reason and faith in history and in the basis of ancient science for the later and modern ages.

Do we even have a place today in this heated debate of whatever may have been or are the powers of the cosmological ancient past? By existential fact of reason we should, but that is a bestial, or primordial, assumption in itself. And the answer would require the attainment of some useful treatise and decision of the otherwise Hermetic topics. So the more basic topical, however mundane, introduction to the topic of Occult Science with Naturalism, needs to be reviewed with greater definition before any relevance of Cabala seems to solve our place with ancient problems, forbidding the complete and personal inspiration of students to be determined and convinced of the entire wit of the Natural argument unified with the Cabala. I prefer not to be the one to suggest such abandon on scientific basis and otherwise would suggest that persons use their religious texts rather than scientific texts to the end of belief. On that account, I enter into the controversies as agreed of controversy and concern for our age.

So even as for instance, there remains to us an age old problem of ciphering hosts from idols and identities even in the influential archaeologies of our rich latter days of an ancient inheritance, we need to gird ourselves with armor as the text here says; and thereafter might be taken up, the unbelievable in the extension of a total faith from an unknown scientific motive to make it seem plausible somehow; since first we should decide of the weather and if it is good enough to venture on that wild journey into lost metaphysical realms, which could only mean void.

That might include, for many people, a more determined interpretation of symbols of the kind as shown here, the Talisman shown here, drawing up the ending of this epilogue a bit more than I have made it already to being loose and transient in order to cover more ground. That the determinism we have given today of our knowledge of science, as for every man to enjoy his physical union to the universe on all physical bases, for himself, is so enjoyable is constantly threatened by the outer domain of knowledge and its esoteric magic; what we learn as children we discard as irrational and that lesson left unsung, returns to us again demanding a statement of rationality due and usually this consults our practices of idols in the wit of our reason and faith together; the meaning of our symbolisms in our cultural and personal usage of symbols. But that is only one example of the esoteric attack on aging human reason. What is it that we know in childhood which avoids the intellectual struggle? To trust in a higher faith of any physical body such as a parent and, or other authority(ies); so much for that in time, that we have forgotten to believe or to trust any authority of rationale, there being a better path of unknown wilderness and bestiality which is the actual hidden confrontation – the cosmological unknown; the union of ourselves to the outer regions of space and substance.

So we need to tackle the magical domain at some point as adults, whether the reasons of childhood exclude the rationale for such a decision to take hold of our interests to do it, by wit of some disappointment or whether the rationale of childhood tells us to try harder by wit of some promise at the outset, which then is forgotten.

For the question will forever make the patient ache in a hospice bed of mysteries to the human soul. And to have reminded us all to childhood again, does that seem unkind, well then, looking around the hermetic table of our knighthood to transcend into the annals of time travels into an ancient understanding, then I have been unkind, because the entire lesson I provoke is to be a knight of armor; and not to disappoint the place of ourselves in childhood thereby, but to preserve our times with greater talents of adult achievements; as to dream with greater depth and purpose to reveal a purpose.

On that note, I will leave it an exercise to everyman to look up the online encyclopedias and find out for himself and herself that what I have asserted is possibly true about this Talisman. It is in itself a great controversy, since I have already gone on at length on the better topic of the book here itself.

FIN

Book 1 Natural and Occult Magic by Francis Barret, 1801

With a Minor Gloss by Patricia Spencer, text editor, 2018

This book might be considered a companion to the title work, Doctrina Antiqua (ed. Patricia Spencer), by Thomas Burnet.

Other antiquated titles in the area of Learned Magic are recommended as ~ (and also by searching the general topics of "Esoteric" and "Grimoires" with relation to library texts of the proven antiquities on the topic of written treatises of magic and cosmogony).

The Book of Sacred Magic by MacGregor Mathers, 1900

The Book of Ceremonial Magic by Arthur Edward Wait, 1913

Three Books of Occult Philosophy by Cornelius Agrippa, 1531-33

I hope to complete at least the first book of Agrippa's Natural Philosophy in this three sequence litany for publication, in the next years soon to appear in similar to this and before any secondary material on the Cabala appears separately presented by myself. However, if students of this kind of literature, are determined to study that section of magic known as Cabala, the recommendation is in the fellowship of those topics noted, also as Ceremonial Magic, Natural Philosophy and Occult Science; besides Esoteric and Grimoires.

Spitzer Telescope ~ The Milky Way Galaxy

Mayan Flat Earth in the Intra-Galactic Cosmos

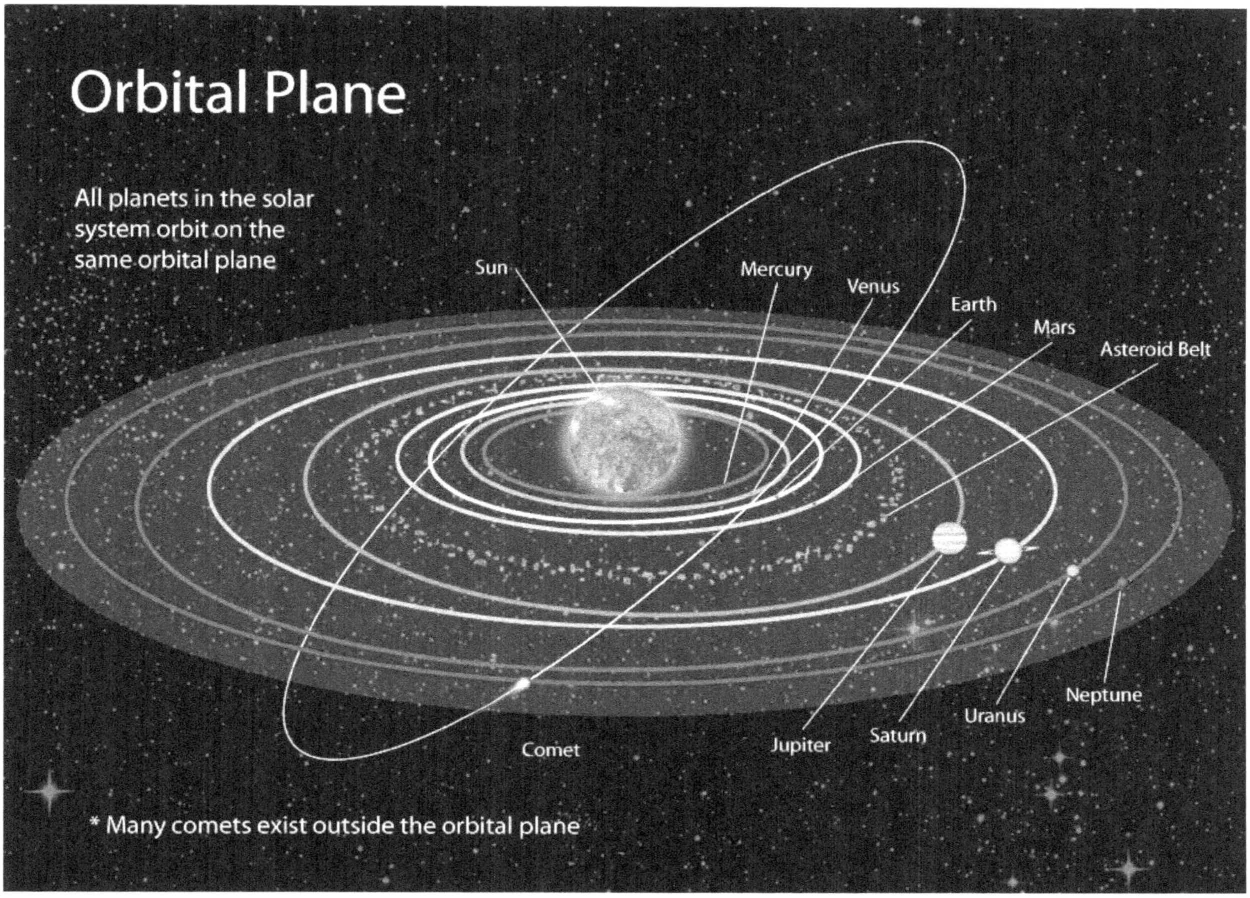

Orbital Plane

All planets in the solar system orbit on the same orbital plane

Sun, Mercury, Venus, Earth, Mars, Asteroid Belt, Jupiter, Saturn, Uranus, Neptune, Comet

* Many comets exist outside the orbital plane

347

Planet X on Outer Pluto ~ aka. Dune or Narragansett

Outer Space ~ aka. the Inner System of the Sun

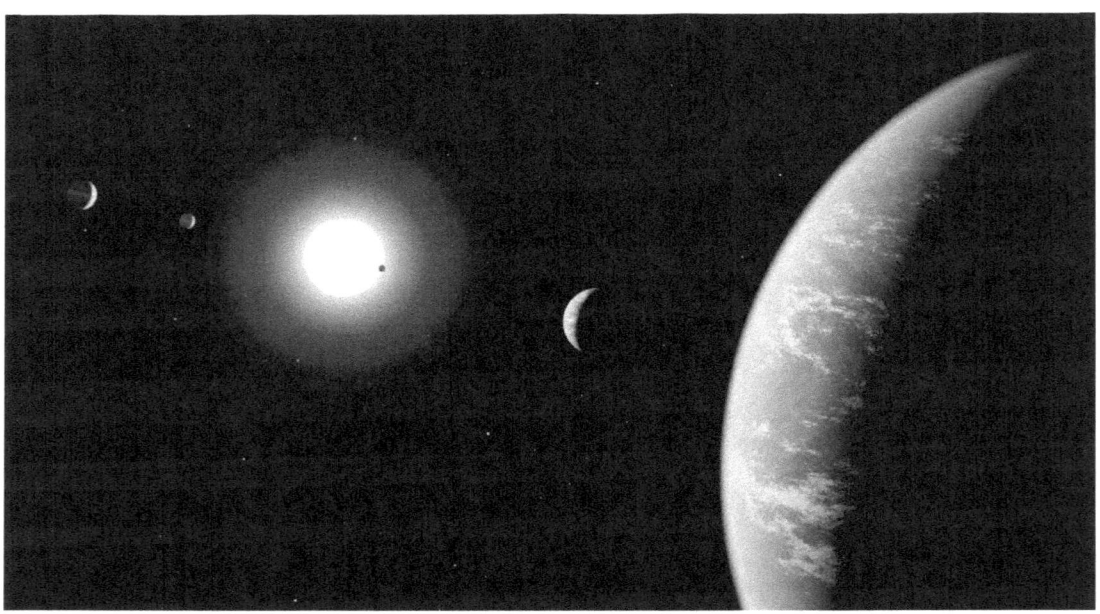

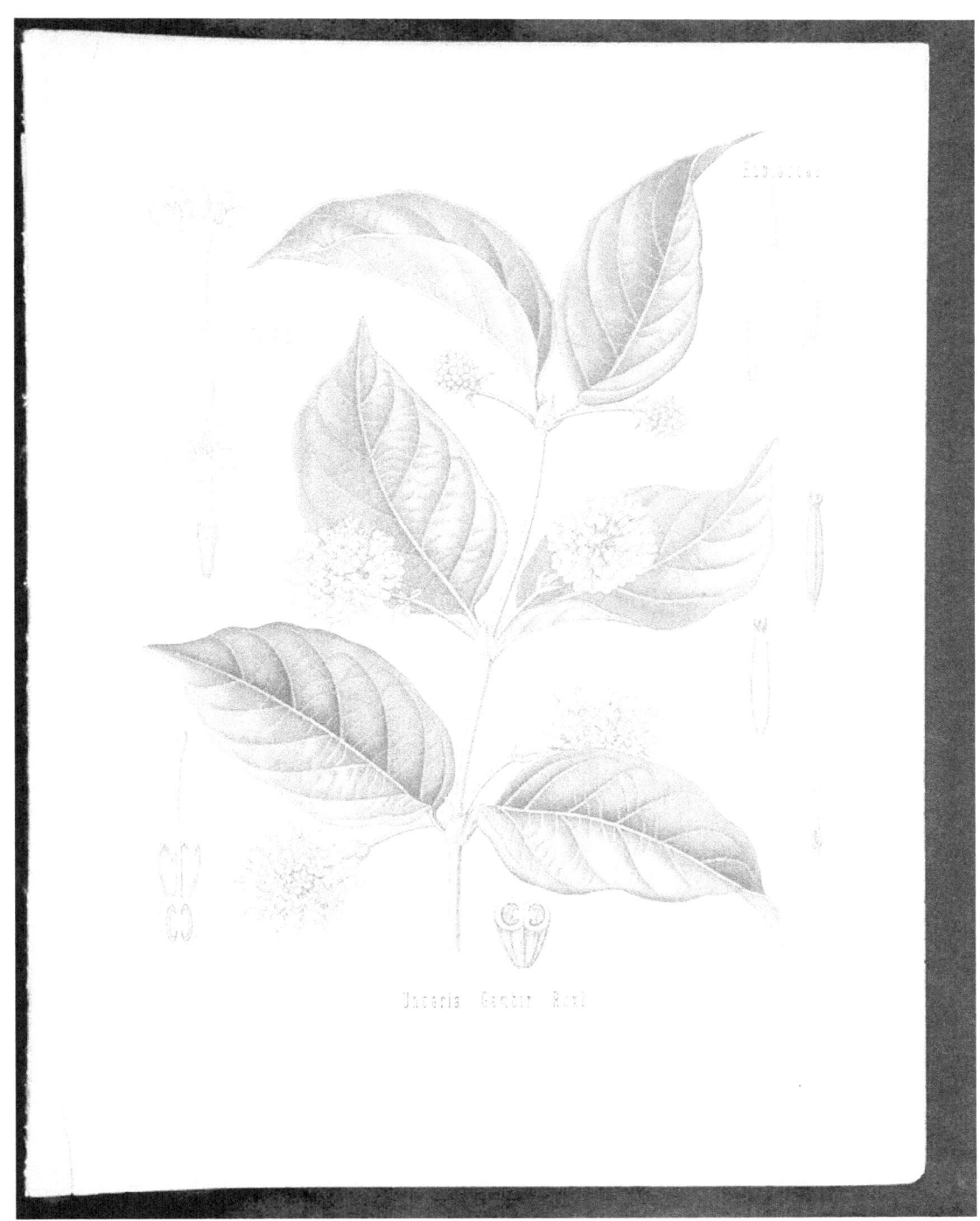

(Uncaria Gambir (Gambier), nicknamed, Amelia Tannin Barkspore)

(Mandrake)

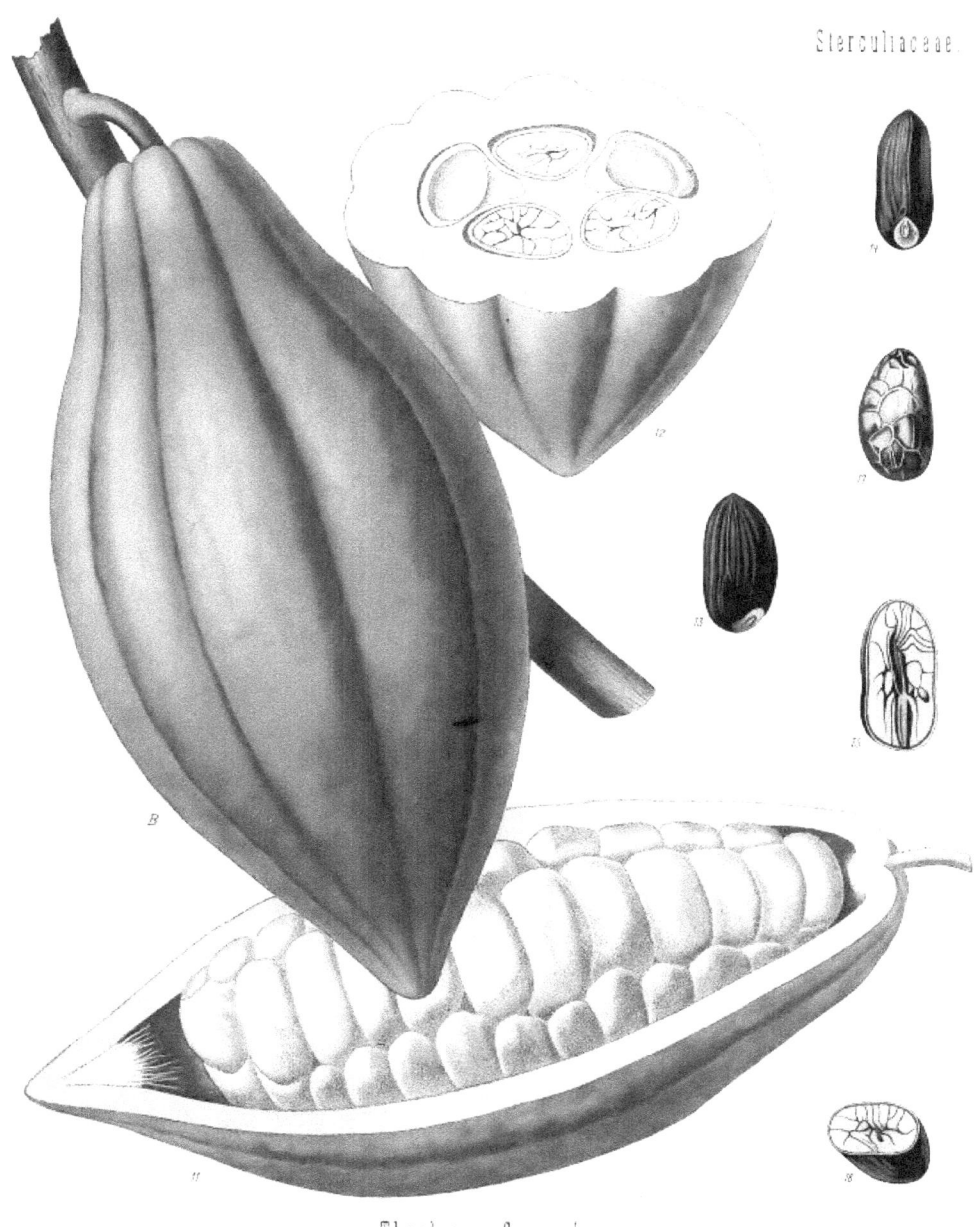

(Tropical, Evergreen Cocoa Tree, also nicknamed, Colonel Coco)

DU GLOBE TERRESTRE.

FIGURE LXVII.

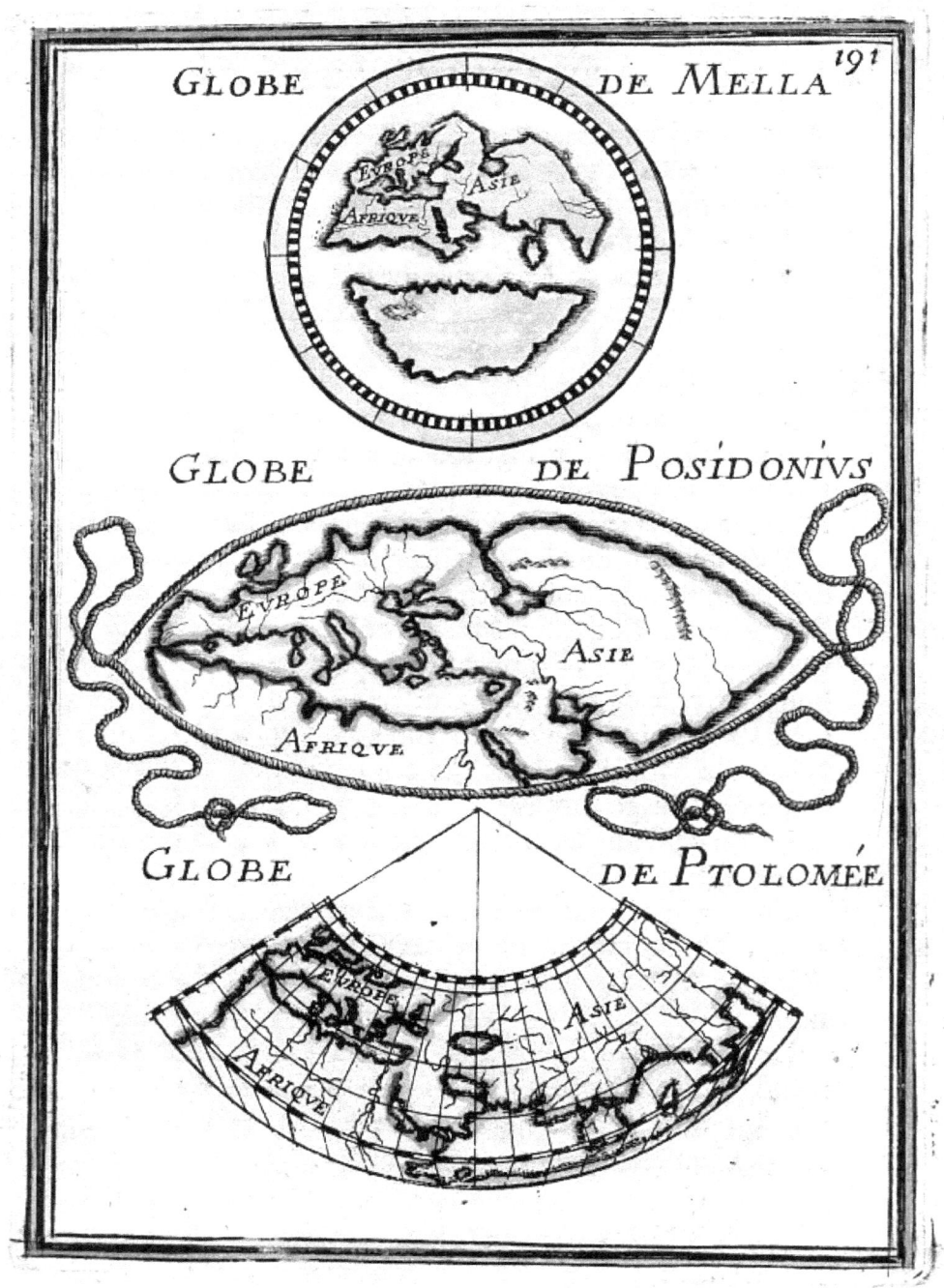

(Moon Dragon, REX Draconis)

A Beach Somewhere in the Universe of Dark Stars

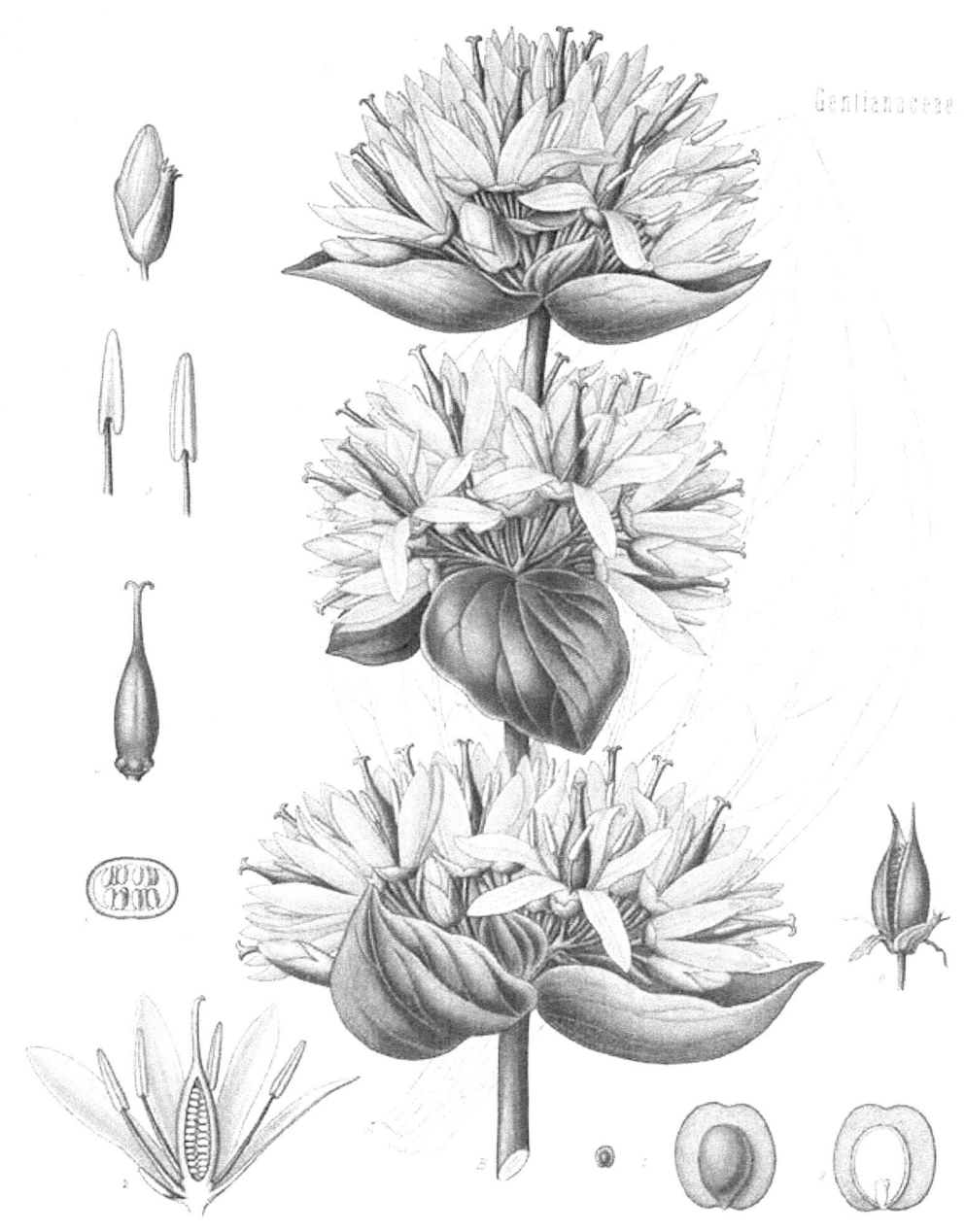

(A Medicinal Plant of name, Bitter Wort,
nicknamed here Cornelia Sorrelswort)

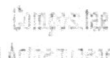

(Another look at Common Mugwort, also nicknamed here
Leafy Spore Wort, formerly Chicory Sorrelspore)

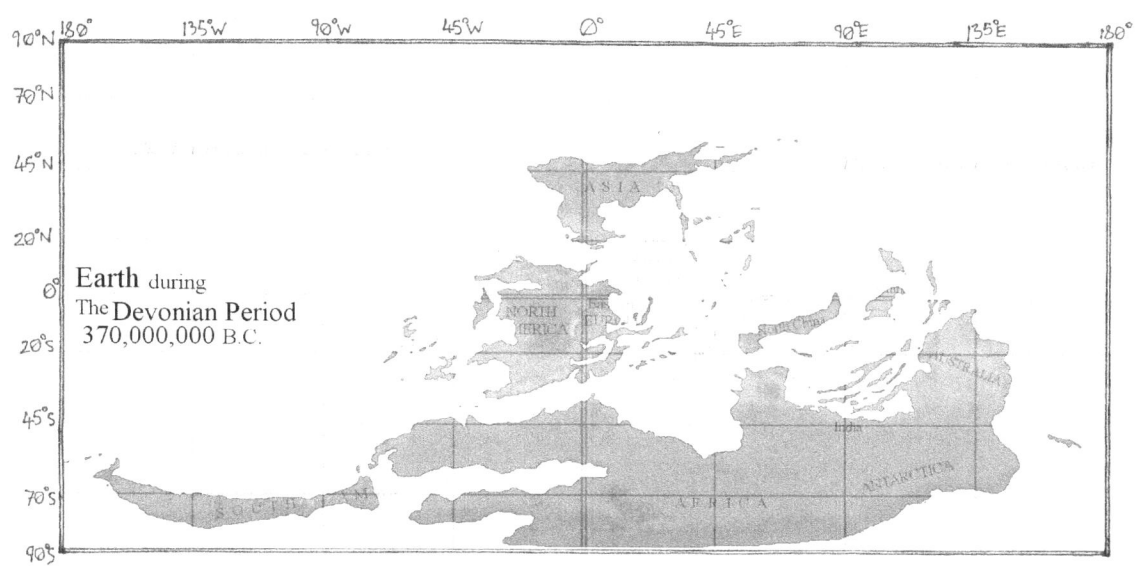

Earth During the Devonian Period 370 million BC

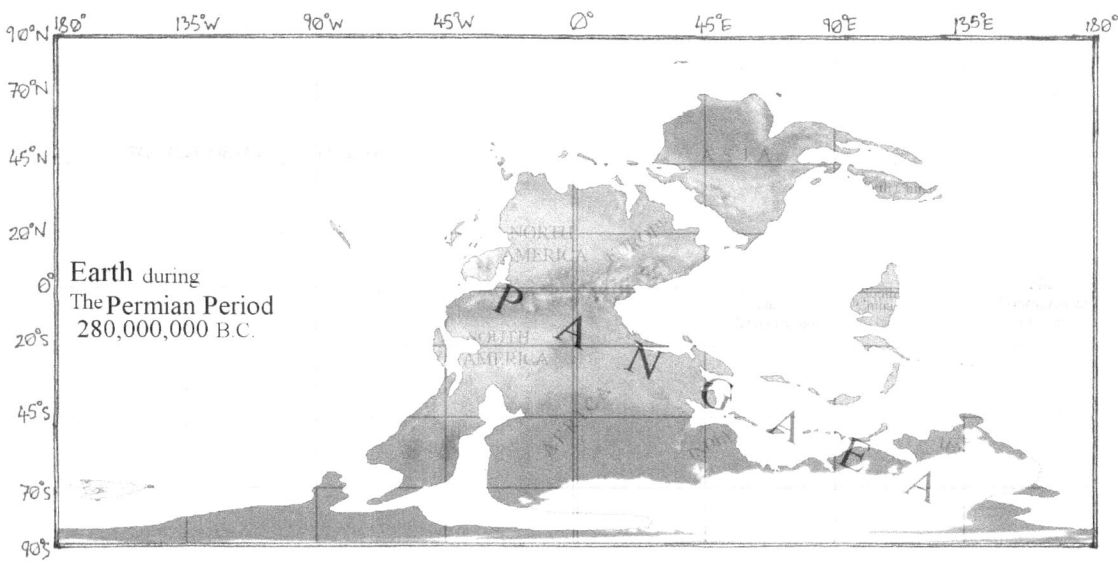

Earth During the Permian Period 280 million BC

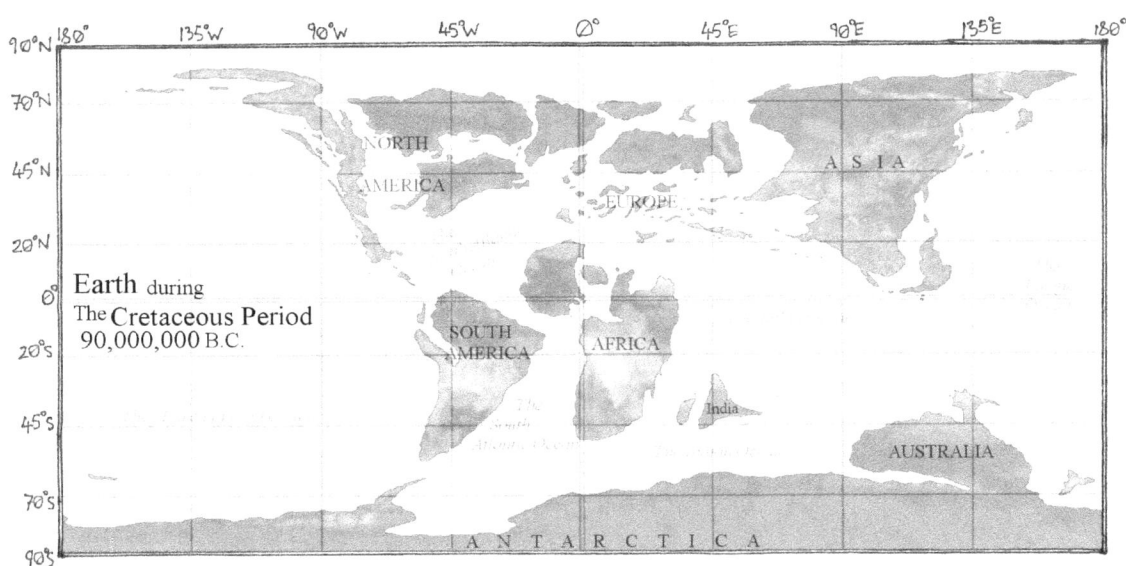

Earth During the Cretaceous Period 90 million BC

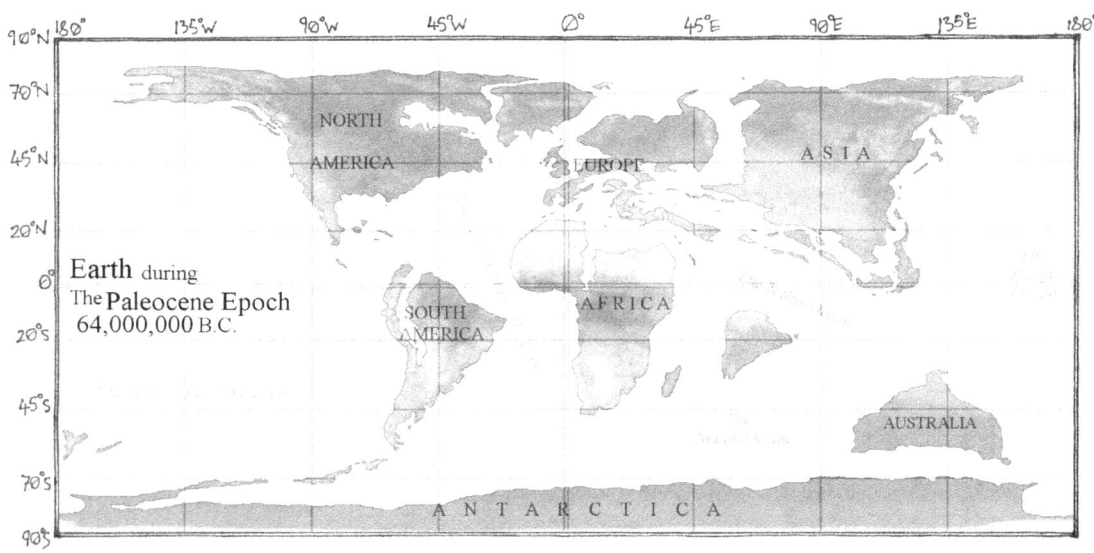

Earth During the Paleocene Epoch 64 million BC

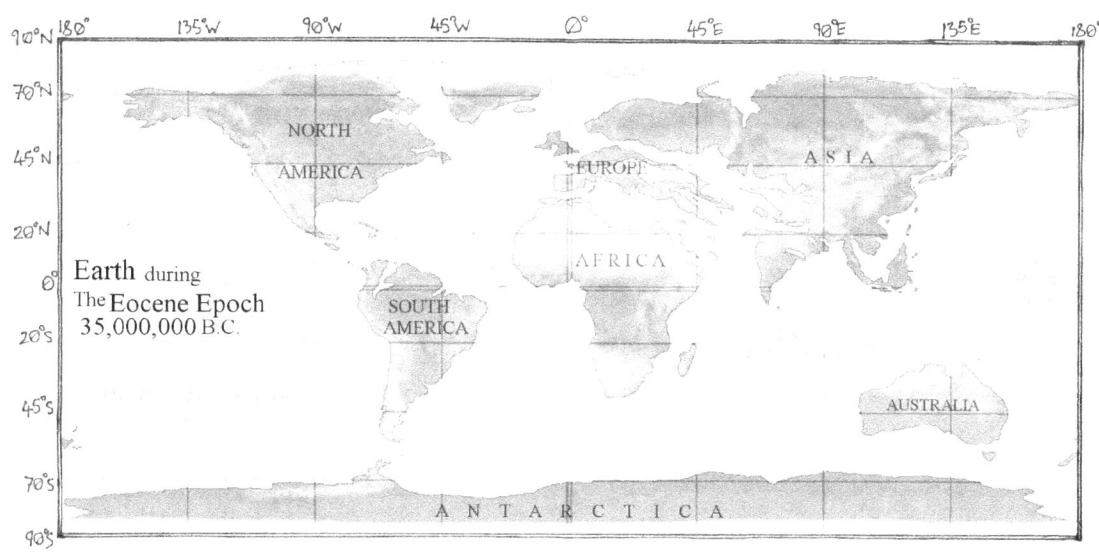

Earth During the Eocene Epoch 35 million BC

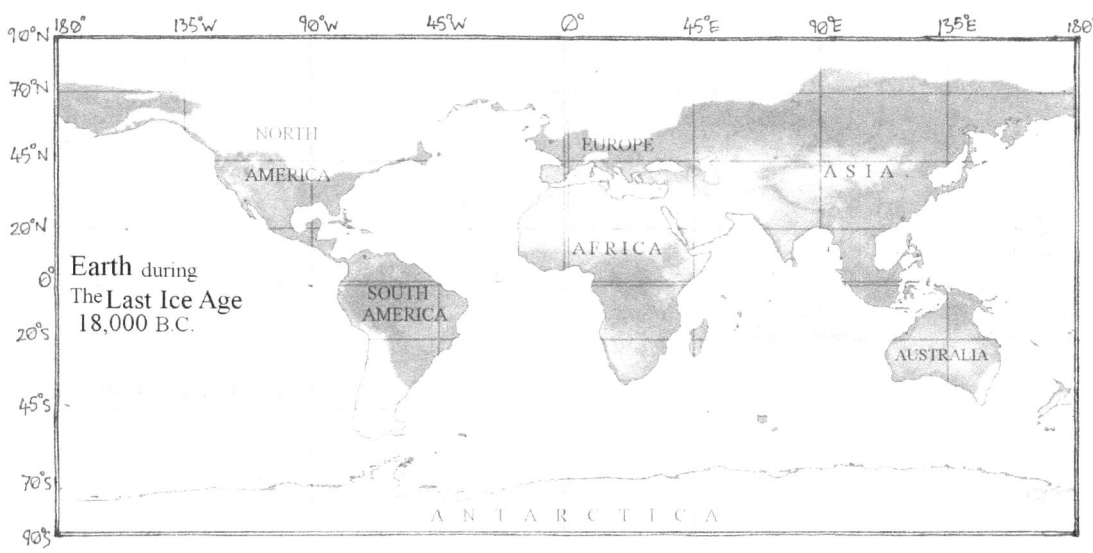

Earth During the Last Ice Age 18 million BC

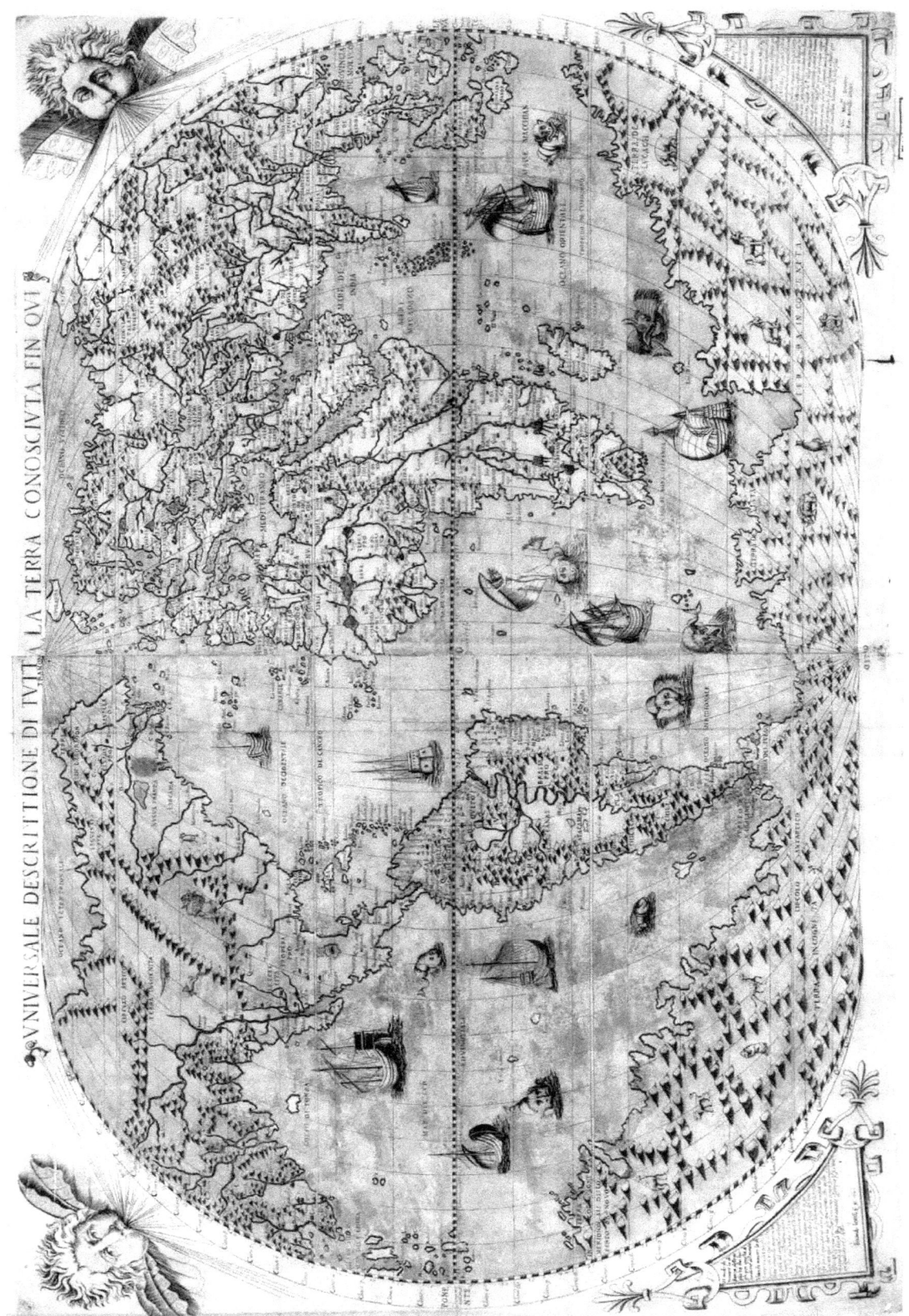

www.ingramcontent.com/pod-product-compliance
Lightning Source LLC
Chambersburg PA
CBHW062212220526
45471CB00009B/3164